Columbia University

Contributions to Education

Teachers College Series

No. 146

AMS PRESS
NEW YORK

SOLVING GEOMETRIC ORIGINALS

BY

FRANK CHARLES TOUTON, Ph.D.

Professor of Education

University of Southern California

Teachers College, Columbia University

Contributions to Education, No. 146

Published by

Teachers College, Columbia University

New York City

1924

Library of Congress Cataloging in Publication Data

Touton, Frank Charles, 1880-1936.
Solving geometric originals.

Reprint of the 1924 ed., issued in series: Teachers College, Columbia University. Contributions to education, no. 146.
Originally presented as the author's thesis, Columbia.
1. Geometry, Plane--Study and teaching. I. Title. II. Series: Columbia University. Teachers College. Contributions to education, no. 146.
QA461.T68 1972 516'.22'07 76-177698
ISBN 0-404-55146-7

Reprinted by Special Arrangement with Teachers College Press, New York, New York

From the edition of 1924, New York
First AMS edition published in 1972
Manufactured in the United States

AMS PRESS, INC.
NEW YORK, N. Y. 10003

PREFACE

Certain phases of this study were suggested to the writer through reading Dewey's *How We Think* and through subsequent analyses of the thought processes utilized in his own solutions of geometric originals. In this study an attempt has been made to determine, if possible, the general method of procedure to be employed in the solution of geometric originals.

Another phase of the study was suggested by the common practice of many teachers, which consists in hearing and correcting demonstrations during the class periods. This practice predominates to the almost total exclusion of the directing of the thinking of pupils in their first attempts at the solution of new and involved geometric situations. In this study answers were sought to the questions raised as to the need for and extent of directed thinking as the proper class room activity in the teaching of geometry. Again the writer has long held a questioning attitude toward the opinion held by many, that girls are inferior to boys in geometric ability. Here, too, an answer was sought in the study. The above considerations have been potent in suggesting and determining the trends which the investigation has taken.

To Professor Thomas H. Briggs, the writer is indebted for suggestions leading up to the selection of the topic, for assistance in securing access to the source materials of the study, and for numerous suggestions given while the study was in progress. To Dr. Charles F. Wheelock and to Mr. George M. Wiley, the writer is indebted for permission to use the papers written by New York State high school pupils in their Regents Examination in Plane Geometry of June, 1918. To Professor Truman Lee Kelley, the writer is indebted for suggestions as to the statistical treatment and for stimulating suggestions as to possible interpretations of the data secured from the papers examined. To Professor David Eugene Smith, the writer is indebted for numerous suggestions made after a careful reading of an early draft of the study. To several able assistants but especially to Miss

Florence E. Paine the writer is indebted for much painstaking and arduous work done in tracing and recording the geometric facts and relations used by the pupils in the solutions which they submitted.

FRANK CHARLES TOUTON

May, 1919

CONTENTS

CHAPTER I

INTRODUCTION

CHAPTER II

THE RELATIVE DIFFICULTIES OF THE EXERCISES

CHAPTER III

SEX DIFFERENCES IN GEOMETRIC ABILITIES CONTRASTED AND CORRELATED

CHAPTER IV

STEPS IN THE THOUGHT PROCESSES INVOLVED IN SOLVING ORIGINALS

CHAPTER V

SUGGESTIONS FOR IMPROVED METHODS OF TEACHING GEOMETRY

CHAPTER VI

RESUMÉ OF THE STUDY

APPENDIX

COMPUTATION OF MEASURES

TABLES

DIAGRAMS

SOLVING GEOMETRIC ORIGINALS

CHAPTER I

INTRODUCTION

Purpose of the Study. The study is based upon a critical analysis of the work done by pupils in their plane geometry examinations with a view to offering such suggestions as the data warrant for improved methods of teaching that subject. As the study progressed data were found, collected, and interpreted which limit its scope to the following main considerations:

1. The difficulties inherent in the several exercises of the examinations.
2. The differences between boys and girls in their choice of certain exercises for solution and in the abilities shown in the solution of those exercises which they select.
3. The success of the pupil in solving originals, as conditioned by his adherence or lack of adherence to an orderly procedure in thinking.
4. Some changes in the teaching methods employed as essential in securing better results from the teaching of plane geometry.

Nature of Materials Used. Through the courtesy of Dr. Charles F. Wheelock, Mr. G. M. Wiley, and Mr. E. E. Arnold, of the New York State Department of Education, the privilege was extended of using approximately 2,800 examination papers written by New York State high school pupils in the New York Regents Examination in Plane Geometry of June 20, 1918.

The high schools from which papers were selected were the following: Akron; Albany: High School and Christian Brothers Academy; Amsterdam; Baldwin; Brooklyn: Boys' High School, Brooklyn Evening School, Commercial High School, Girls' High School, Manual Training High School, New Utrecht, and Saint Francis Academy; Canajoharie; Canton; Carthage; Cathedral High School; Chatham; Cooperstown; Cortland; Dansville; Dobbs Ferry; Drum Hill; East Aurora; Ellenville; Fairport;

Fayetteville; Fulton; Geneva; Glens Falls; Gouverneur; Haverling; Herkimer; Holley; Ilion; Ithaca; Johnstown; Kingston; Lackawanna; Lansingburg; Little Falls; Lockport; Lynbrook; Massena; Medina; Middletown; Monticello; Newark; Newburgh; New Rochelle; New York City: Bryant, DeWitt Clinton, Newtown, Julia Richman, and Wadleigh; North Tonawanda; Norwich; Norwood; Nyack; Oneida; Orchard Park; Palmyra; Penn Yan; Plattsburg; Port Jervis; Poughkeepsie; Ripley; Rockville Center; South Side High School; Roxbury; Salamanca; Saranac Lake; Savannah; Schenectady; St. Joseph's Academy; Sidney; South Hampton; Syracuse Central; Ticonderoga; Troy; Walton; Waterford; Waterville; Wellsville; Yonkers: High School and Seaton Academy.

In this study no records were kept which could be interpreted in such a way as to provide a reliable contrast of the several schools as to the efficiency of instruction in geomerty. Had the scores of pupils been classified by schools, even then no reliable contrast of instruction would have been possible, for the mental ability scores of the pupil-groups had not been determined.

Since it is not the practice of New York State high school teachers to submit to the Department of Education at Albany papers of pupils which do not in the judgment of the teachers possess sufficient worth to secure a passing grade on the examination as a whole, since the pupils were allowed to make certain selections from among the exercises of the list, and since the quality of the work done on each exercise is considered by itself in the study, it is believed that the work done shows fewer low scores than would be shown had all the papers written by the pupils in these schools been read and reported upon.

From the character of the schools in the above list and from the selection of papers as explained above, it is probable that the quality of work examined was somewhat above the average of that done by all pupils of the state of New York. We may assume, therefore, that such conclusions as are drawn in the study are based upon an examination of the work of a selected group of high school pupils.

The examination, as set, consists of thirteen questions, the first four of which require the reproduction of proofs of theorems and the last nine of which demand original work on the part of the pupils. These exercises are reproduced on pages 3 and 4.

EXERCISES 5–13 INCLUSIVE, FROM THE NEW YORK REGENTS EXAMINATION IN PLANE GEOMETRY, JUNE 20, 1918

5. The area of a square inscribed in a circle is 16 square inches. Find the area of the circle.

6. The radius of a circle is 15 inches. Through a point 5 inches from the center a chord is drawn. What is the product of the two segments of the chord? What is the length of the shortest chord that can be drawn through that point?

7. The base of a triangle is 20 feet; the other sides are 10 and 16 feet. A line parallel to the base cuts off 2 feet from the lower end of the shorter side. Find the segments of the other side and the length of the parallel.

8. Construct a line tangent to a given circle and parallel to a given line outside the circle. (To receive credit construction lines must be shown.)

9. A. An exterior angle of a regular polygon equals 1/5 of a right angle. Find the number of sides of the polygon.

   B. The difference between two angles inscribed in the same circle is 20°. What is the difference between the two central angles subtended by the arcs of the inscribed angles?

   C. If the side of one equilateral triangle equals the altitude of another, what is the ratio of their areas?

   D. The sides of a triangle are 3 feet and 8 feet. What are the numerical limits of the third side?

10. ABC is a triangle. D is the foot of the perpendicular from A on BC, P is the middle point of BC, X is a point on BC such that $XP = PD$. If the line through P perpendicular to BC meets AX at M, prove that $MB = MC$ and $MX = MA$.

11. From the external point P, a secant PM is drawn to a circle so that it is bisected by the circle at N. MD is the diameter through M. Prove that PD is equal to the diameter and state a simple method for drawing a secant from an external point to a circle so that it will be bisected by the circle.

12. Prove that a line perpendicular to a side of a right triangle at its middle point passes through the mid-point of the hypotenuse.

13. *ABC* is a triangle and *AD, BE,* and *CF* are its medians. *DH* is drawn equal and parallel to *BE,* and cuts *AC.* Prove that *HA* is equal and parallel to *CF.*

Before beginning the study of the work done by pupils, a tentative list was made of several possible solutions of the foregoing exercises with the essential steps involved in each such solution. Then through a close analysis of the work of the pupils, it was possible to record not only the successes or failures of the pupils but also the facts actually used by them and also the frequency with which each fact was called into use. This analysis of the work done by pupils is reported on the following pages with tables showing the use made by pupils of certain geometric facts. It is upon this record of the work done by pupils and upon the grades assigned by the teachers and examiners (see page 106) that the findings and suggestions of the study are based.

Analysis of Pupils' Work. Each essential fact necessary to a correct solution is scored as: (1) correct, (2) unsatisfactory, (3) not used, or (4) wrongly used. By *correct use* is meant a fact correctly stated and properly used. By an *unsatisfactory use* is meant an imperfect or inadequate use which indicated at least some familiarity with the necessary data on the part of the pupil. *None* or *not used* designates a total lack of knowledge of the fact on the part of the pupil as evidenced by his failure to state or use the fact at the proper point. By *wrong use* is meant an incorrect statement, or an incorrect use of a fact, or its use in the wrong place.

A brief survey of the following analyses with appended data, showing the ways in which pupils reacted, indicates that there are wide individual differences among pupils who have pursued the same subject, in responding to the same geometric situations. An examination of these data must convince any teacher that it is useless to expect even a large percentage of pupils to respond in any given way to an original situation. What these differences among pupils are, how they may be accounted for, and how methods of teaching should be changed because of these observed differences are among the subjects investigated and reported upon in this study.

Scope of the Study. Upon the basis of these investigations, certain recommendations are made, both for teaching plane geometry and for measuring the results of such teaching. The study will treat in order the following questions:

1. What differences in difficulty are inherent in the problems?

2. What differences in difficulty are experienced by boys and girls in solving certain problems?

3. What does the study show as to the thought processes used by those pupils who react most successfully in their solution of original exercises?

4. What methods of teaching plane geometry are suggested by the study?

Exercise 5

The area of a square inscribed in a circle is 16 square inches. Find the area of the circle.

Facts and Principles Used or Necessary to Method Adopted

(1) Concept of a square. (Sides =, and ∠ s all ∠ rts.)
(2) Concept of the area of a square as measured in square units.
(3) Concept of a circle.
(4) Concept of a figure inscribed in a circle.
(5) The area of the circle is to be expressed in square units.
(6) Sketch representing the given conditions.
(7) Area of a square in terms of side s is s^2; here $s^2 = 16$. Whence $s = 4$.
(8) $(\text{Diagonal})^2$ or $(\text{Diameter})^2 = 4^2 + 4^2 = 32$. Whence $d = 4\sqrt{2}$ and $r = 2\sqrt{2}$.
(9) To find the area of the circle the radius must first be found.
(10) In an inscribed square the diagonals bisect each other at the center of circle. (Statement was not specifically required.)
(11) Area of circle $= \pi r^2 = \pi(2\sqrt{2})^2 = 8\pi = 25\ 1/7$ or $8 \times 3.1416 = 25.1328$.
(12) Extraction of square root.
(13) Purely numerical errors affecting results (addition, subtraction, multiplication and division).
(14) Errors in solving equations.
(15) Used 1–7 above and 10. Then used $r^2 + r^2 = 4^2 = 16$. Whence $r = 2\sqrt{2}$.
(16) Used 1–8 above and 10. Then used $s : 4 = 4 : s/2$.
(17) Used 1–8 above and 10. Then used $d \times d/2 = 16$.
(18) Used 1–10 above and used: Area $= (2 \times \pi \times r) \times r/2$ or $\pi d \times r/2 = 8\pi = 25\ 1/7$ or 25.1328.
(19) Used 1–8 above and 10. Then used $(s/2)^2 + (s/2)^2 = r^2$ or $2^2 + 2^2 = r^2$. Whence $r = 2\sqrt{2}$.
(20) Area $= 1/4\ \pi d^2$.
(21) The area of a circle equals π times the square of the radius.
(22) In a right triangle the square of the hypotenuse equals sum of squares of legs.
(23) The area of a circle equals one half the circumference times the radius.
(24) The diagonals of a square are equal and bisect at right angles.
(25) The area of a square equals one half the product of its diagonals.
(26) The circumference of circle equals $2\pi r$ or πd.
(27) In a circle equal chords are equally distant from the center.
(28) The apothem of a square equals one half its side.
(29) Diagonals cut the square into 4 equilateral triangles. Then used area $= s^2\sqrt{3/4}$.
(30) Angles inscribed in a semicircle are ∠ rts. (Used to prove ∠ s of square, ∠ rts.)
(31) In a rt △ either leg is a mean proportional between hypotenuse and adjacent segment.
(32) In a circle a diameter perpendicular to a chord bisects the chord, etc.
(33) An incorrect guess or statement made at random, apparently.
(34) Area of circle equals product of circumference and diameter, πd. (Statement incorrect.)

Record on Use Made of Facts

	Total		*Correct*		*Unsatisfactory*		*None*		*Wrong*	
	Boys	*Girls*	*Boys*	*Girls*	*Boys*	*Girls*	*Boys*	*Girls*	*Boys*	*Girls*
(1)	541	607	504	511	37	91	0	5	0	0
(2)	546	607	414	482	0	0	128	124	4	1
(3)	546	607	543	603	3	1	0	3	0	0
(4)	546	607	544	597	1	6	1	3	0	1
(5)	546	607	263	240	0	0	273	354	10	13
(6)	546	596	499	486	38	99	7	9	2	2
(7)	540	605	436	470	49	73	20	23	35	39
(8)	387	420	340	361	16	15	10	11	21	33
(9)	544	603	525	562	2	5	13	23	4	13
(10)	547	606	12	10	535	596	0	0	0	0
(11)	555	584	385	348	81	140	11	6	78	90
(12)	194	288	163	252	16	19	0	0	15	17
(13)	79	90	0	0	0	0	0	0	79	90
(14)	21	38	0	0	0	0	0	0	21	38
(15)	132	99	122	92	5	0	2	0	3	7
(16)	3	11	2	10	1	0	0	0	0	1
(17)	10	18	8	16	0	0	0	0	2	2
(18)	10	13	4	4	0	6	1	0	5	3
(19)	34	39	31	36	1	0	0	2	2	1
(20)	54	99	45	90	0	9	0	0	9	0
(21)	540	602	506	573	2	5	16	8	16	16
(22)	533	586	494	532	5	2	29	46	5	6
(23)	38	9	29	7	0	1	3	0	6	1
(24)	151	105	115	97	34	5	2	0	0	3
(25)	10	44	9	23	0	8	0	13	1	0
(26)	6	9	6	8	0	0	0	0	0	1
(27)	27	28	27	28	0	0	0	0	0	0
(28)	30	32	30	29	0	0	0	0	0	3
(29)	3	6	0	0	0	0	0	0	3	6
(30)	0	2	0	1	0	0	0	0	0	1
(31)	2	11	2	10	0	0	0	0	0	1
(32)	26	27	26	27	0	0	0	0	0	0
(33)	3	14	0	0	0	0	0	0	3	14
(34)	6	6	0	0	0	0	0	0	6	6

Exercise 6

The radius of a circle is 15 inches. Through a point 5 inches from the center a chord is drawn. What is the product of the two segments of the chord? What is the length of the shortest chord that can be drawn through that point?

Facts and Principles Used or Necessary to Method Adopted

(1) Concept of a circle.
(2) Concept of the radius of a circle.
(3) Concept of a chord.
(4) Concept of segments of a chord made by a point on it.
(5) Distance from the point to the center is the distance measured along a diameter or radius through that point.
(6) The point cuts the diameter through it into segments 10 and 20.
(7) Product of the segments of the chord is 200.
(8) Answers in *square inches.*
(9) Found product of segments of shorter chord *only.*
(10) Theorem used: If two chords are drawn through a point within a circle, the product of the segments of one chord equals the product of the segments of other.
(11) Absurd error or statement without any apparent reason.
(12) Method not evident.
(13) Distance of a point to a line is distance along a $\perp$ from the point to the line.
(14) Shortest chord will be twice the length of the base of a right $\triangle$, whose hypotenuse is the radius, and whose other side is the distance of the chord from the center of the circle.
(15) One half the chord $= \sqrt{15^2 - 5^2} = \sqrt{200} = 10\sqrt{2}.$
(16) Chord equals $20\sqrt{2}$ or 28.28.
(17) Error in use of radicals.
(18) Error in extracting square root.
(19) Use of general theorem A $(\text{chord}/2)^2 = 20 \times 10$ or chord $= \sqrt{800}.$
(20) $20 : c/2 = c/2 : 10$. Whence $c = 20\sqrt{2} =$ etc.
(21) Answer expressed in inches.
(22) Error through use of $(\text{chord}/2)^2 = 15^2 + 5^2$.
(22) Error through use of $(\text{chord}/2)^2 = 15^2 + 5^2$, etc.
(24) Used $x : 20 = 10 : x$, etc.
(25) In a circle the shorter of two chords is farther from center.
(26) In a right triangle or semicircle the perpendicular is a mean proportional between the segments of the hypotenuse or diameter.
(27) In a right triangle the square of one leg equals the square of the hypotenuse minus the square of the other leg.
(28) The diameter perpendicular to a chord bisects the chord.
(29) Error through use of $30 : x = x : 10$.
(30) Used fact that: In similar triangles homologous sides are proportional.
(31) Numerical errors.

RECORD ON USE MADE OF FACTS

	Total		*Correct*		*Unsatisfactory*		*None*		*Wrong*	
	Boys	*Girls*	*Boys*	*Girls*	*Boys*	*Girls*	*Boys*	*Girls*	*Boys*	*Girls*
(1)	522	528	521	528	0	0	1	0	0	0
(2)	522	528	519	528	1	0	1	0	1	0
(3)	522	528	508	522	11	4	2	0	1	2
(4)	521	528	435	432	34	37	45	54	7	5
(5)	521	527	519	524	0	0	2	3	0	0
(6)	521	526	243	146	13	7	252	364	13	9
(7)	512	518	361	314	15	25	56	107	80	72
(8)	522	528	39	58	19	25	404	373	60	72
(9)	187	272	0	0	0	0	0	0	187	272
(10)	518	527	244	137	9	8	265	380	0	2
(11)	1	3	0	0	0	0	0	0	1	3
(12)	5	3	0	0	0	0	5	3	0	0
(13)	521	524	515	515	0	0	5	9	1	3
(14)	458	468	390	408	9	14	47	34	12	12
(15)	452	470	340	362	18	16	46	46	48	46
(16)	492	515	304	306	58	58	49	50	81	101
(17)	20	25	0	0	0	0	0	0	20	25
(18)	28	31	0	0	0	0	0	0	28	31
(19)	55	25	42	22	1	0	0	0	12	3
(20)	26	29	21	22	3	5	0	1	2	1
(21)	517	522	139	145	0	0	372	372	6	5
(22)	11	5	0	0	0	0	0	0	11	5
(23)	9	12	0	0	0	0	0	0	9	12
(24)	9	14	6	9	0	0	1	0	2	5
(25)	520	525	499	507	0	1	19	15	1	2
(26)	21	30	18	27	1	0	2	1	0	2
(27)	445	467	391	424	2	2	49	39	3	2
(28)	519	512	491	481	0	8	26	20	2	3
(29)	1	4	0	0	0	0	0	0	1	4
(30)	8	8	5	5	0	1	0	0	3	2
(31)	30	54	0	0	0	0	0	0	30	54

Exercise 7

The base of a triangle is 20 feet; the other sides are 10 and 16 feet. A line parallel to the base cuts off 2 feet from the lower end of the shorter side. Find the segments of the other side and the length of the parallel.

Facts and Principles Used or Necessary to Method Adopted

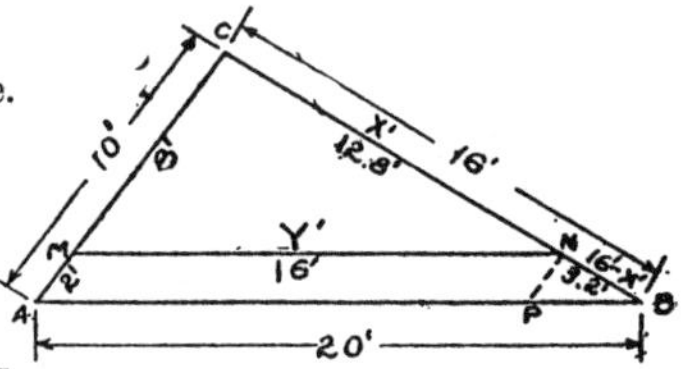

(1) Concept of a triangle.
(2) Concept of parallel lines.
(3) Meaning of a segment or segments of a line.
(4) Use of 2 feet for the lower segment of the shorter side.
(5) Use 20 feet or the base as a horizontal line.
(6) The *two* segments of the longer side are required.
(7) The length of the line || to the base is required.
(8) Sketch of a figure representing the given data.
(9) The upper segment of the shorter side is 8 feet.
(10) The segments of the 16-foot side made by the || to the base are x and $16 - x$.
(11) $8 : 2 = x : 16 - x$.
(12) $128 - 8x = 2x$.
(13) $x = 12.8$ and $16 - x = 3.2$.
(14) (1) $8 : 10 = y$ (the parallel) $: 20$ or (2) $12\,4/5 : 16 = y : 20$.
(15) $160 = 10y$ or $y = 16$.
(16) Segments of longer side and the parallel are expressed in feet.
(17) $10 : 8 = 16 : x$. Whence, $10x = 128$, $x = 12.8$, $16 - x = 3.2$.
(18) $16 : 20 = x : 16$. Whence, $x = 12.8$, $16 - x = 3.2$.
(19) A line || to the base of a △ divides the adjacent sides into proportional segments.
(20) A line || to the base of a △ cuts off a △ similar to the whole △.
(21) In similar △s homologous sides are proportional.
(22) In a proportion the product of the means equals the product of the extremes.
(23) Used vertex angle as a ∠ rt. Special case.
(24) Homologous sides occupy corresponding positions in similar triangles.
(25) Drew || to *MA* through *N*, saw parallelogram, used sim. △s, and found *AP* or y.
(26) Absurd statement or method not evident.
(27) Parallel to base = ½ of base, therefore, $y = 10$. (Incorrect.)
(28) Drew || to *MA* through *N*, used theorem in (19). (Incorrect.)
(29) $10 : 2 = 16 : x$, etc.
(30) $10x = 32$, $x = 3.2$ and $16 - x = 12.8$.
(31) One side of a △ is to base segment of it made by || as other side is to corresponding base segment.
(32) Product of means = product of extremes in a proportion.

Record on Use Made of Facts

	Total		*Correct*		*Unsatisfactory*		*None*		*Wrong*	
	Boys	*Girls*	*Boys*	*Girls*	*Boys*	*Girls*	*Boys*	*Girls*	*Boys*	*Girls*
(1)	499	625	499	625	0	0	0	0	0	J
(2)	499	625	499	625	0	0	0	0	0	J
(3)	499	625	431	594	68	29	0	0	0	2
(4)	498	626	492	624	1	0	1	1	4	1
(5)	499	625	499	618	0	5	0	0	0	2
(6)	499	615	421	526	45	30	30	58	3	1
(7)	499	625	446	527	0	1	50	89	3	8
(8)	499	625	497	623	0	0	0	3	2	1
(9)	499	626	482	601	1	1	9	19	7	2
(10)	499	623	463	579	22	18	4	13	10	13
(11)	218	192	118	132	2	7	1	4	97	49
(12)	135	165	126	135	2	5	0	4	7	21
(13)	258	207	194	129	47	34	0	5	17	39
(14)	485	604	374	411	4	29	69	120	38	44
(15)	479	604	363	398	9	8	49	112	58	86
(16)	495	610	143	171	24	20	322	415	6	4
(17)	129	138	117	107	5	17	1	9	6	5
(18)	10	33	8	30	1	3	0	0	1	0
(19)	213	199	201	166	0	2	4	12	8	19
(20)	490	606	421	473	5	13	55	108	9	12
(21)	487	621	421	484	5	13	56	112	5	12
(22)	492	610	491	590	0	0	1	18	0	2
(23)	5	8	0	0	0	0	0	0	5	8
(24)	488	621	389	444	6	8	57	114	36	55
(25)	7	9	5	8	0	0	0	0	2	1
(26)	1	0	0	0	0	0	0	0	1	0
(27)	5	4	0	0	0	0	0	0	5	4
(28)	2	1	0	0	0	0	0	0	2	1
(29)	146	275	140	267	1	6	1	0	3	2
(30)	114	275	100	201	38	68	1	6	5	10
(31)	145	283	142	278	1	5	0	0	2	0
(32)	145	286	145	285	0	0	0	1	0	0

Exercise 8

Construct a line tangent to a given circle and parallel to a given line outside the circle. (*To receive credit construction lines must be shown.*)

Facts and Principles Used or Necessary to Method Adopted

(1) To construct means to draw accurately by means of a straight edge and a pair of compasses.
(2) A straight line must be drawn.
(3) Concept of a circle.
(4) Concept of a tangent to a circle.
(5) Fact that the given line must not intersect the circle.
(6) Concept of parallel lines.

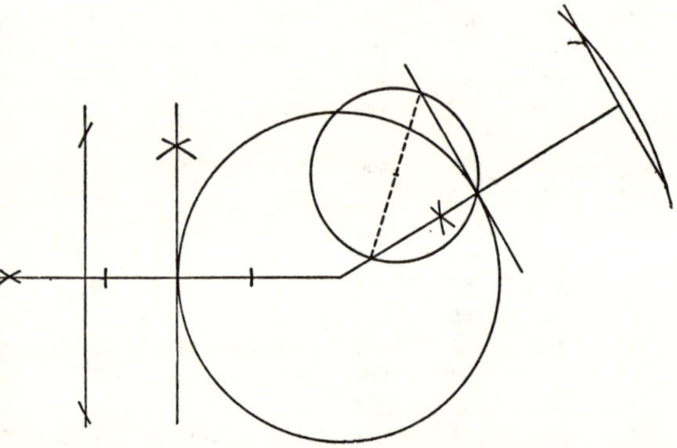

(7) The given line and the given circle must be selected before the construction of the required line is begun.
(8) Drawing of a line from the center of the given circle perpendicular to the given line, showing construction lines.
(9) Selection of the intersection of the perpendicular and the circumference as the point through which the tangent to the circle must be drawn.
(10) Construction of a perpendicular to the constructed line at the point where that line cuts the circumference, using the mid-perpendicular construction.
(11) Construction lines must all be clearly shown.
(12) Two perpendiculars to the same straight line are parallel.
(13) A tangent to a circle is perpendicular to the radius at the point of contact.
(14) Two points each equally distant from the extremities of a line determine a mid $\perp$.
(15) Two lines are $\parallel$ if (a) corresponding $\angle$ s are $=$. (b) if alternate int. $\angle$ s are $=$.
(16) Use 1–9 above and construct a $\perp$ to the radius at end using $\angle$ ins. in semicircle.
(17) Use 1–9 above and then (a) construct corresponding $\angle$ rts, or (b) alt. int. $\angle$ rts.
(18) Perpendicular distances between $\parallel$ s are $=$. Corresponding construction of tangent.
(19) Unnecessary construction lines.
(20) Construction correct if line outside is not considered given but may be constructed.
(21) Use 1–9 above, construct $\angle$ rt outside figures, cons. its equal to locate tangent.
(22) Construction correct for a line intersecting the circle.
(23) Construction correct if circle is not considered given, but may be constructed.

Record on Use Made of Facts

	Total		*Correct*		*Unsatisfactory*		*None*		*Wrong*	
	Boys	*Girls*	*Boys*	*Girls*	*Boys*	*Girls*	*Boys*	*Girls*	*Boys*	*Girls*
(1)	503	579	503	578	0	0	0	0	0	1
(2)	503	579	503	579	0	0	0	0	0	0
(3)	503	579	503	579	0	0	0	0	0	0
(4)	503	579	501	578	1	1	0	0	1	0
(5)	502	579	496	561	0	0	1	5	5	12
(6)	503	579	500	574	0	3	2	2	1	0
(7)	503	578	455	446	0	0	48	132	0	0
(8)	493	562	388	317	32	71	66	161	7	13
(9)	493	554	429	384	1	2	61	167	2	1
(10)	377	398	281	205	2	1	89	191	5	1
(11)	502	568	445	440	53	118	4	10	0	0
(12)	501	565	453	458	2	5	46	102	0	0
(13)	499	569	475	521	6	11	18	37	0	0
(14)	487	531	433	460	1	2	53	69	0	0
(15)	80	223	76	216	1	3	3	4	0	0
(16)	107	76	106	71	1	5	0	0	0	0
(17)	47	123	45	108	1	1	0	0	1	14
(18)	9	7	7	5	2	2	0	0	0	0
(19)	45	105	0	0	0	0	0	0	45	104
(20)	35	96	33	86	0	5	0	0	2	5
(21)	0	1	0	1	0	0	0	0	0	0
(22)	1	6	1	4	0	2	0	0	0	0
(23)	7	23	7	23	0	0	0	0	0	0

EXERCISE 9

A. An exterior angle of a regular polygon equals 1/5 of a right angle. Find the number of sides of the polygon.

FACTS AND PRINCIPLES USED OR NECESSARY TO METHOD ADOPTED

(1) Concept of an exterior angle.
(2) Fact that the number of angles equals the number of sides.
(3) Concept of a regular polygon as having equal exterior angles.
(4) The sum of the exterior angles of a polygon equals 360°.
(5) $1/5$ of $90° = 18°$.
(6) $360° \div 18° = 20$, or the number of sides.
(7) $180 - 18 = 162$. $n(n-2)/n \times 180 = 162$. Whence, $n = 20$.
(8) 4 rt. $\angle$s $\div$ 1/5 $\angle$ rts $= 20$.
(9) Interior $\angle = 9/5 \angle$ rts $= 2([n-2]/n)$ rt. $\angle$s. Whence, $n = 20$.
(10) Central $\angle = 18°$, therefore 20 sides.

B. The difference between two angles inscribed in the same circle is 20°. What is the difference between the two central angles subtended by the arcs of the inscribed angles?

(1) Concept of an inscribed angle.
(2) Concept of a central angle.
(3) Concept of an arc subtending an angle.
(4) An inscribed angle is measured by ½ its intercepted arc.
(5) A central angle is measured by its intercepted arc.
(6) The central angle equals two times the inscribed angle intercepting same arc.
(7) The difference of the inscribed angles is 20°.
(8) The difference of the central angles is 40°.
(9) Used a special case to get conclusions.
(10) Took difference to mean difference in position.

C. If the side of one equilateral triangle equals the altitude of another, what is the ratio of their areas?

(1) Concept of an equilateral triangle.
(2) Concept of an altitude.
(3) Concept of ratio.
(4) In an equilateral $\triangle$ of sides s and altitude a, $a = s/2\sqrt{3}$.
(5) Altitudes are to each other as $s : s/2\sqrt{3}$.
(6) Areas are to each other as $(s)^2 : (s/2\sqrt{3^2})$.
(7) Areas of equilateral triangles are to each other as the squares of altitudes or sides.
(8) Areas are to each other as $3:4$ or as $4:3$.
(9) Used a special case.
(10) Areas are to each other as $hb/2 : h'b'/2 = hb/h'b' = b^2 : b'^2$.

(Continued on Pages 16 and 17)

RECORD ON USE MADE OF FACTS

	Total		*Correct*		*Unsatisfactory*		*None*		*Wrong*	
	Boys	*Girls*	*Boys*	*Girls*	*Boys*	*Girls*	*Boys*	*Girls*	*Boys*	*Girls*
(1)	258	243	225	200	0	0	33	43	0	0
(2)	258	242	218	188	0	0	39	54	1	0
(3)	258	243	220	187	0	0	38	56	0	0
(4)	242	233	187	159	0	0	40	68	15	5
(5)	250	234	207	186	0	1	37	39	6	8
(6)	237	221	172	139	1	2	33	50	31	30
(7)	15	21	9	12	3	5	0	0	3	4
(8)	6	8	3	8	1	0	0	0	2	0
(9)	3	1	2	1	1	0	0	0	0	0
(10)	0	1	0	1	0	0	0	0	0	0
(1)	258	238	246	216	0	0	12	20	0	2
(2)	258	243	246	213	0	0	12	29	0	1
(3)	258	223	243	196	0	0	15	26	0	1
(4)	258	243	234	203	0	0	22	38	2	2
(5)	258	242	235	196	0	0	22	43	1	3
(6)	252	241	212	177	1	0	22	44	17	20
(7)	253	241	235	203	0	1	16	34	2	3
(8)	256	240	215	170	0	0	13	39	28	31
(9)	27	4	18	4	0	0	0	0	9	4
(10)	0	1	0	0	0	0	0	0	0	1
(1)	258	243	138	154	1	3	118	86	1	0
(2)	258	243	139	154	1	1	117	88	1	0
(3)	257	243	161	162	1	1	95	80	0	0
(4)	257	243	46	40	5	5	195	198	11	0
(5)	257	243	24	20	1	2	229	221	3	0
(6)	224	243	19	23	2	4	193	208	10	8
(7)	258	243	55	34	4	14	168	172	31	23
(8)	258	242	39	8	2	6	172	186	45	42
(9)	11	8	0	0	0	0	0	0	11	8
(10)	24	38	13	31	3	1	0	0	8	6

Exercise 9—(*continued*)

D. The sides of a △ are 3 and 8 feet. What are the numerical limits of the third side?

Facts and Principles Used or Necessary to Method Adopted

(1) Meaning of numerical limits.
(2) Third side must be greater than 5 feet.
(3) Third side must be less than 11 feet.
(4) Found limits of angle between given sides to nearest minute, 1′–179° 59′.
(5) Not more than 11 or less than 5 feet.
(6) Answer, 5. No reason given. Geometrical concept of a limit.
(7) Answer, 11. No reason given. Geometrical concept of a limit.
(8) Used rt △, found hypotenuse = $\sqrt{73}$. Stated $\sqrt{73}$ as the answer.

RECORD ON USE MADE OF FACTS

	Total		*Correct*		*Unsatisfactory*		*None*		*Wrong*	
	Boys	*Girls*	*Boys*	*Girls*	*Boys*	*Girls*	*Boys*	*Girls*	*Boys*	*Girls*
(1)	142	183	28	16	11	15	83	121	20	31
(2)	159	135	40	21	23	5	71	96	25	13
(3)	154	132	93	51	9	5	42	61	10	15
(4)	1	0	1	0	0	0	0	0	0	0
(5)	5	8	0	0	5	8	0	0	0	0
(6)	23	15	16	10	0	0	5	3	2	2
(7)	26	15	22	10	0	0	2	3	2	2
(8)	4	5	0	0	0	0	0	0	4	5

Exercise 10

ABC is a triangle. D is the foot of the perpendicular from A on BC; P is the middle point of BC; X is a point on BC such that XP equals PD. If the line through P perpendicular to BC meets AX at M, prove that MB = MC and MX = MA.

Facts and Principles Used or Necessary to Method Adopted

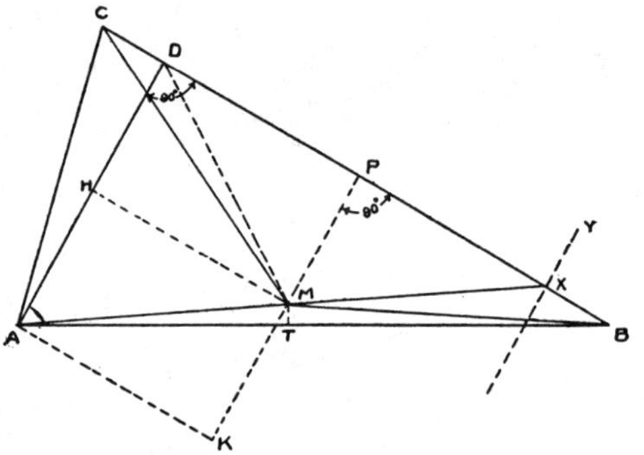

(1) Concept of a triangle.
(2) Concept of the foot of a perpendicular.
(3) Sketched a figure for conditions given.
(4) Given: △ *ABC*.
(5) Given: *AD* ⊥ *BC* at *D*.
(6) Given: *CP* = *PB*.
(7) Given: *XP* = *PD*.
(8) Given: *PM* ⊥ to *CB* at *P*.
(9) To prove *MB* = *MC*, and *MX* = *MA*.
(10) *MB* = *MC*. If a ⊥ is erected at the mid-point of a line, any point in that ⊥ is equally distant from the extremities of the line.
(11) In △s *MBP* and *MPC*, (1) leg and leg, or (2) *S* ∠ S. Hence △s =. *MB* = *MC*. Homologous parts of = △s =.
(12) In △ *ADX*, *PM* is ∥ to *AD*.
(13) Lines are ∥, (1) ∥ to the same straight line. (2) cor. ∠s. (3) alt. int. ∠s.
(14) *PX* = *PD*. Given.
(15) Then *MX* = *MA*.
(16) A line ∥ to the base of a △ and bisecting one side bisects the other also.
(17) △s *MPX* and *ADX* are sim. (∠ ∠ ∠). *AX*:*MX*=*DX*:*PX* = 2:1. *AM* = *MX*.
(18) Homologous sides of sim. △s are proportional.
(19) *DX* = 2*PX*. Given. *AX* = 2*MX*. From above, hence *AM* = *MX*.
(20) Drew *MH* ⊥ *AD*. Proved △s *XPM* and *DPM* = and △s *DMH* and *MHA* =. Then *MA* = *MX*. (Homologous parts.)
(21) Assumed ∠ *XMP* = ∠ *DMB*, then proved △s =, then *XM* = *MD*.
(22) Three or more ∥s cutting off = distances on a trans., etc.
(23) Random statements, unrelated or inaccurate.
(24) Drew *XY* ⊥ *BC* or ∥ to *AD*. *XY*, *MP*, and *AD* are ∥. Since *DP* = *PX*, *AM* = *MX*. Three ∥s, etc.
(25) Proved *MX* = *MD*. *MD* = *AM*. Then *MX* = *MA*. Used cor. ∠s theorem, alt. int. ∠s theorem in proving △s similar.
(26) *DP*: *PX* = *MA*: *MX*. Since *DP* = *PX*, *MA* = *MX*.
(27) Line ∥ to base of △ divides sides proportionately, etc.
(28) Drew from *M* a ⊥ to *AD*. Showed *AM* = *MB*: Used rt. △s *MBP* and *PMX*.
(29) Drew *MH* ⊥ *AD* and bisecting it.
(30) Considered ∠ *AMB* and ∠ *MBC* as rt. ∠s.
(31) Produced *PM*. Drew *AK* ⊥ *PM*. Proved *AKM* = *MPX*. (∠ *S* ∠). *AM* =*MX*. (Hom. parts.)

RECORD ON USE MADE OF FACTS

	Total		*Correct*		*Unsatisfactory*		*None*		*Wrong*	
	Boys	*Girls*	*Boys*	*Girls*	*Boys*	*Girls*	*Boys*	*Girls*	*Boys*	*Girls*
(1)	463	552	462	552	0	0	1	0	0	0
(2)	463	552	462	552	0	0	1	0	0	0
(3)	463	552	450	526	12	25	0	0	1	1
(4)	463	552	457	543	1	1	5	7	0	1
(5)	463	552	457	542	0	0	5	10	1	0
(6)	463	551	458	543	0	0	5	8	0	0
(7)	463	552	456	541	1	0	5	11	1	0
(8)	463	552	457	540	1	0	4	11	1	1
(9)	463	551	455	543	3	2	5	6	0	0
(10)	173	155	162	146	7	8	2	1	2	0
(11)	291	394	257	345	19	33	5	10	10	6
(12)	452	526	283	239	1	2	168	285	0	0
(13)	451	527	245	216	4	8	196	297	6	6
(14)	453	535	299	266	2	1	152	268	0	0
(15)	449	535	300	255	0	0	149	275	0	0
(16)	289	371	117	84	13	9	156	271	3	7
(17)	10	11	9	3	1	7	0	0	0	1
(18)	9	7	9	6	0	0	0	0	0	1
(19)	6	1	3	0	2	1	0	0	1	0
(20)	16	13	9	5	5	6	0	0	2	2
(21)	0	4	0	0	0	0	0	0	0	4
(22)	67	61	43	18	22	41	1	0	1	2
(23)	17	27	0	0	0	0	0	0	17	27
(24)	54	30	46	26	7	4	1	0	0	0
(25)	26	28	15	12	9	9	0	0	2	7
(26)	31	51	15	29	13	19	1	3	2	0
(27)	27	48	27	44	0	4	0	0	0	0
(28)	0	4	0	2	0	1	0	0	0	1
(29)	1	3	0	0	0	0	0	0	1	3
(30)	1	8	0	0	0	0	0	0	1	8
(31)	11	6	7	3	3	2	0	0	1	1

Exercise 11

From the external point P, a secant PM is drawn to a circle so that it is bisected by the circle at N. MD is the diameter through M. Prove that PD is equal to the diameter and state a simple method for drawing a secant from an external point to a circle so that it will be bisected by the circle.

Facts and Principles Used or Necessary to Method Adopted

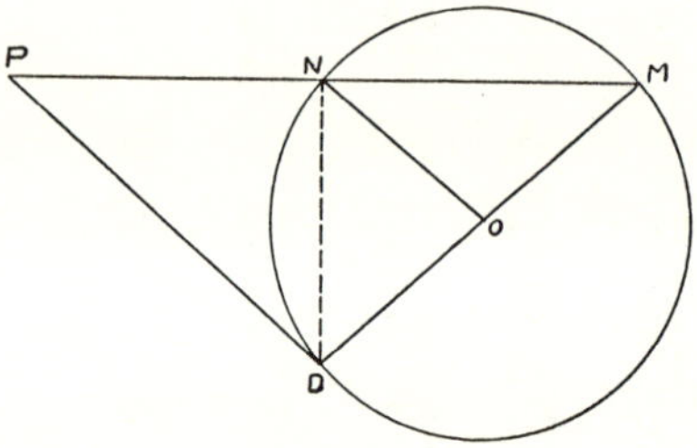

(1) Meaning of an external point.
(2) Concept of a secant.
(3) Concept of a circle.
(4) Concept of a diameter.
(5) Meaning of bisected.
(6) Sketch of the above conditions.
(7) Given circle with center O.
(8) Given secant PM with $PN = NM$.
(9) Given diameter MD.
(10) To prove: $MD = PD$.
(11) Draw chord DN.
(12) $\angle DNM$ is a $\angle$rt.
(13) An $\angle$ inscribed in semicircle is a $\angle$rt.
(14) Since $\angle DNM$ is a $\angle$rt, then $\angle DNP$ is a $\angle$rt.
(15) Definition of a perpendicular or
(16) Adjacent $\angle$s having exterior sides in same straight line are supplementary.
(17) $DN = DN$. Identity.
(18) $PN = NM$. Given.
(19) $\triangle$s DNM and DPN are equal. Two sides and included angle, etc., or
(20) Right $\triangle$s having leg and leg equal respectively.
(21) Therefore $PD = DM$. Homologous sides of equal triangles are equal.
(22) With the given point P as a center and with a radius equal to the diameter of the given circle strike an arc intersecting the circumference.
Draw the diameter of the circle through this point. Connect the other extremity of the diameter with point P. Thus is the required secant drawn.
(23) Noted that distance from point to O must not be greater than diameter through O.
(24) Used 1–10 above and drew radius ON.
(25) ON equals $\frac{1}{2}PD$, since $OM = OD$ (radii of circle) and $NP = NM$. (Given.)
(26) A line connecting the midpoints of two sides of a $\triangle$ equals $\frac{1}{2}$ the base.
(27) $PD = 2NO = MD$, hence $PD = MD$. (Equality axiom.)
(28) Used 1–16 above. Also $PN = NM$. (Given.)
(29) DN is perpendicular to MP. Lines meet at a $\angle$ rt.
(30) $DP = DM$. Point in mid perpendicular is equally distant from extremities of line.
(31) Therefore theorem or Q.E.D.
(32) $\angle P = \angle M$. Both measured by $\frac{1}{2}$ arc ND. Therefore PMD is an isosceles $\triangle$.

RECORD ON USE MADE OF FACTS

	Total		*Correct*		*Unsatisfactory*		*None*		*Wrong*	
	Boys	*Girls*	*Boys*	*Girls*	*Boys*	*Girls*	*Boys*	*Girls*	*Boys*	*Girls*
(1)	507	551	504	551	1	0	2	0	0	0
(2)	506	551	497	546	2	1	2	1	5	3
(3)	506	551	505	551	0	0	1	0	0	0
(4)	506	551	504	543	0	7	2	1	0	0
(5)	506	551	500	546	1	2	3	3	2	0
(6)	506	551	489	533	10	6	4	6	3	6
(7)	506	452	338	361	83	18	85	67	0	6
(8)	510	551	425	487	45	27	36	36	4	1
(9)	508	550	457	496	4	2	47	52	0	0
(10)	506	550	476	526	4	0	26	22	0	2
(11)	461	546	338	376	0	7	89	97	34	66
(12)	491	544	401	389	10	12	76	140	4	3
(13)	502	544	390	355	10	9	89	164	3	16
(14)	489	535	349	349	23	30	113	148	4	8
(15)	128	70	76	42	17	3	33	20	2	5
(16)	303	466	128	140	4	111	164	210	7	5
(17)	483	514	433	455	3	2	44	54	3	3
(18)	463	515	415	461	11	15	33	38	4	1
(19)	381	418	301	343	3	3	33	25	44	47
(20)	77	99	62	74	0	6	4	4	11	15
(21)	428	509	389	458	2	3	28	39	9	9
(22)	507	551	16	11	79	76	251	266	161	198
(23)	25	4	6	3	0	0	0	0	19	1
(24)	12	6	11	5	1	1	0	0	0	0
(25)	14	6	9	4	1	1	2	1	1	0
(26)	13	6	8	3	1	0	4	3	0	0
(27)	13	5	8	3	1	2	4	0	0	0
(28)	30	26	26	23	1	0	3	3	0	0
(29)	39	29	33	21	3	1	3	7	0	0
(30)	36	32	33	27	2	2	0	2	1	1
(31)	33	31	11	18	0	0	22	13	0	0
(32)	2	8	0	2	0	0	0	2	2	4

Exercise 12

Prove that a line perpendicular to a side of a right triangle at its middle point passes through the mid-point of the hypotenuse.

Facts and Principles Used or Necessary to Method Adopted

(1) Concept of perpendicular lines.
(2) Concept of a right angle.
(3) Concept of the mid-point of a line.
(4) Concept of the hypotenuse of a right triangle.
(5) Sketch of a figure to represent the given conditions.
(6) Given: $\angle ABC$ is a right angle.
(7) M is the mid-point of AB.
(8) AC is the hypotenuse.
(9) MN is $\perp$ to AB at M.
(10) To prove $AN = NC$.

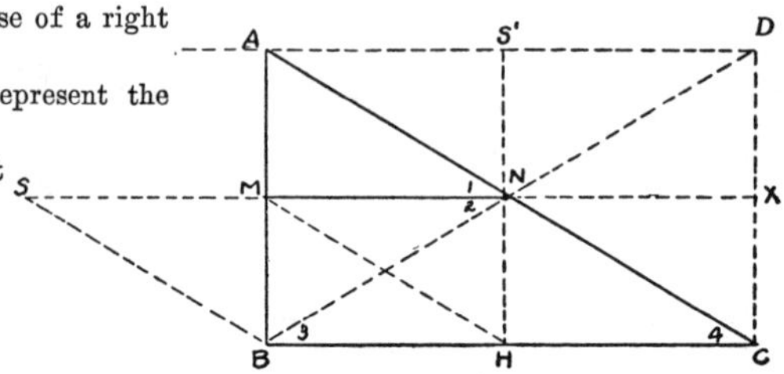

(11) MN and BC are $\parallel$.
(12) Lines $\perp$ to the same straight line are $\parallel$; alt. int. $\angle$s; cor. $\angle$s.
(13) N is the mid-point of AC, MN is $\parallel$ to BC and bisects AB.
(14) A line $\parallel$ to the base of a triangle bisecting one side bisects the other also.
(15) Impossible constructions: $\parallel$ or $\perp$ to a fixed point; $BC \parallel$ to AC; bisector $\angle B$ bisects AC.
(16) $AM : MB = AN : NC$.
(17) A line $\parallel$ to base of $\triangle$ divides the sides proportionally.
(18) Therefore $AN = NC$, for $AM = MB$, or $AM : MB = 1 : 1$, or prod. means = prod. extr. with equals divided by equals gives equals.
(19) Used 1–12 above and drew AD through $A \parallel MN$ (or $\perp$ to AB with 12 above).
(20) AD is $\parallel$ to BC.
(21) A line $\parallel$ to one of two $\parallel$s is $\parallel$ to the other also.
(22) $AN = NC$.
(23) If three $\parallel$s intercept equal segments on one of two transversals, they intercept equal segments on the other also.
(24) Used 1–12 above and drew a $\perp$ NH, from N to BC.
(25) $MB = NH$. ($NH \perp MN$, see 12 above.) Perpendiculars between $\parallel$s are =.
(26) $NH = AM$ (= same thing). $\triangle AMN = \triangle HCN$ (ASA). Then $AN = NC$. (Hom. sides of = $\triangle$s.)
(27) Or used 1–10 above and drew $CX \parallel$ to AB and produced MN to meet CX at X.
(28) CX is $\perp$ to MN produced.
(29) A $\perp$ to one of two $\parallel$s is perpendicular to the other also.
(30) $CX \parallel MB$, for MN and BC are $\parallel$. (Drawn $\parallel$ making $BMXC$ a parallelogram.)
(31) Perpendiculars to the same straight line are parallel.
(32) $CX = BM$. $\parallel$s included between $\parallel$s are =.
(33) $CX = MA$. They each equal BM. Equality axiom.
(34) $\angle ANM = \angle NCH$. (Corr. $\angle$s are =.)

(*Continued on Pages 24 and 25*)

RECORD ON USE MADE OF FACTS

	Total		Correct		Unsatisfactory		None		Wrong	
	Boys	Girls	Boys	Girls	Boys	Girls	Boys	Girls	Boys	Girls
(1)	544	558	542	555	0	1	2	2	0	0
(2)	544	558	543	558	1	0	0	0	0	0
(3)	544	558	543	556	0	2	1	0	0	0
(4)	544	557	542	552	1	2	1	3	0	0
(5)	543	558	537	548	5	3	1	3	0	4
(6)	544	558	526	550	15	3	3	4	0	1
(7)	543	556	526	546	10	0	7	11	0	0
(8)	543	558	532	544	8	11	3	0	0	3
(9)	543	556	530	550	4	0	9	5	0	1
(10)	543	558	537	554	1	0	4	4	1	0
(11)	526	520	447	416	2	0	76	100	1	4
(12)	506	511	429	392	3	1	71	109	3	9
(13)	205	216	177	162	2	2	26	48	0	4
(14)	203	210	174	156	2	4	26	44	1	6
(15)	0	9	0	0	0	0	0	0	0	9
(16)	129	145	121	138	5	7	1	0	2	0
(17)	128	148	124	143	2	3	2	2	0	0
(18)	127	147	114	138	11	6	2	3	0	0
(19)	111	79	79	35	0	1	32	43	0	0
(20)	108	81	53	26	1	0	54	55	0	0
(21)	106	82	53	26	0	0	53	56	0	0
(22)	109	82	108	82	0	0	1	0	0	0
(23)	111	85	96	76	14	8	1	0	0	1
(24)	3	3	3	2	0	0	0	1	0	1
(25)	3	4	3	0	0	0	0	1	0	3
(26)	5	4	5	1	0	0	0	0	0	2
(27)	16	12	13	11	1	1	1	0	1	0
(28)	15	12	12	10	0	1	2	1	1	0
(29)	15	11	10	9	1	1	3	1	1	0
(30)	16	14	12	14	0	0	3	0	1	0
(31)	15	12	12	9	0	1	2	2	1	0
(32)	16	13	14	11	0	0	1	2	1	0
(33)	16	13	13	11	0	0	2	2	1	0
(34)	17	15	15	13	0	0	1	0	1	2

EXERCISE 12—(*continued*)

Prove that a line perpendicular to a side of a right triangle at its middle point passes through the mid-point of the hypotenuse.

FACTS AND PRINCIPLES USED OR NECESSARY TO METHOD ADOPTED

(35) $\triangle AMN = \triangle NXC$. ($\angle S \angle$.)

(36) $AN = NC$. (Homologous parts of $= \triangle$s.)

(37) Circumscribed semi-circle about rt $\triangle$. Connected N and B. Recognized diam. as hypotenuse.

(38) Assumed N the mid-point of AC. (To be proved.)

(39) Drew $CD \parallel$ to AB, and $AD \parallel$ to BC. Proved $ABCD$ a $\parallel$-gram, then a rectangle. BD and AC diags.

(40) The diagonals of a rectangle are = and bisect each other.

(41) Drew $BS \parallel$ to NC, proved $\triangle BMS = \triangle AMN$. (*SAS*.) $BS = AN$. $BCNS$ a $\parallel$-gram, hence $BS = NC$, then

(42) Proved AMN and ABC sim. (Rt. $\triangle$s with = acute $\angle$s.) Then homolog. sides prop., etc.

(43) Used 1–12 above. Then drew $NH \perp BC$ or $\parallel$ to AB. Proved $AN = NC$ from $= \triangle$s. (*ASA*.)

(44) Used 1–10 above. Drew $NH \perp BC$. Drew NB. Proved $\triangle$s $NHC = NHB$; $NMB = ANM$; $AN = NC$.

(45) Used 1–10 above. Proved $NA = NB$ from $= \angle$s NHC and NHB. Proved $NB = \mathrm{NC}$. (Isos. $\triangle$.) $AN = NC$.

(46) Used 1–10, rts. $\angle AMN$ and ABC sim.; then $AB : AM = AC : AN$; $AC = NC$. (See 17 above.)

(47) Used 1–10. Proved rts. $\triangle AMN = ABC$ sim.; homol. sides prop.; $MN \parallel BC$; then 14–15 (see page 22) $AN = NC$.

(48) Proved rts. $\triangle AMN = MNB$; hence $CN = NB$. Proved rt. $\angle BNH = CMH$. Hence $CN = NB$. Hom. parts.

(49) $BN = NC$ for $\angle 1 = \angle 4$, $\angle 2 = \angle 3$, $\angle 1 = \angle 2$, hence $\angle 3 = \angle 4$. $AN = BN = NC$. Proved $BCX = ABH$.

(50) Proved rt. $\triangle AMN =$ rt. $\triangle MNB$, $AN = NB$, but stopped there.

RECORD ON USE MADE OF FACTS

	Total		*Correct*		*Unsatisfactory*		*None*		*Wrong*	
	Boys	*Girls*	*Boys*	*Girls*	*Boys*	*Girls*	*Boys*	*Girls*	*Boys*	*Girls*
(35)	17	13	12	11	0	1	1	0	4	1
(36)	16	12	15	11	0	1	1	0	0	0
(37)	4	2	1	1	0	1	0	0	3	0
(38)	4	4	0	0	0	0	0	0	4	4
(39)	5	4	1	3	1	0	0	0	3	1
(40)	3	1	3	0	0	0	0	0	0	1
(41)	3	3	2	2	1	1	0	0	0	0
(42)	14	33	13	24	1	8	0	0	0	1
(43)	23	7	16	1	3	6	0	0	4	0
(44)	9	19	6	10	3	8	0	0	0	1
(45)	23	34	12	20	10	14	0	0	1	1
(46)	11	11	5	6	4	5	0	0	2	0
(47)	0	1	0	0	0	1	0	0	0	0
(48)	4	7	2	2	2	5	0	0	0	0
(49)	2	2	0	0	0	0	0	0	2	2
(50)	0	2	0	1	0	0	0	0	0	1

Exercise 13

ABC is a triangle and AD, BE, and CF are its medians. DH is drawn equal and parallel to BE, and cuts AC. Prove that HA is equal and parallel to CF.

Facts and Principles Used or Necessary to Method Adopted

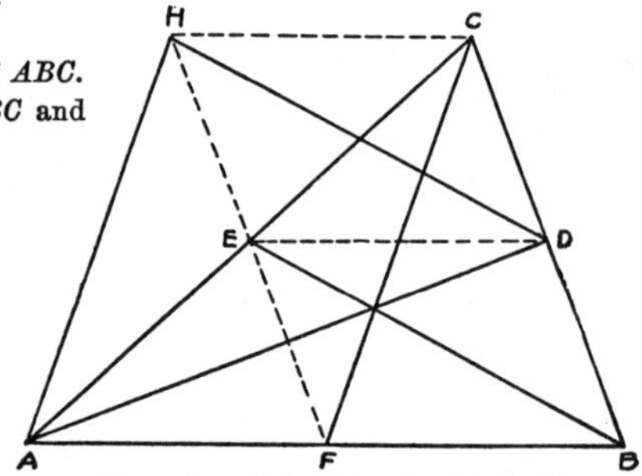

(1) Concept of a triangle.
(2) Concept of a median of a triangle.
(3) Concept of parallel lines.
(4) Given: *BE, CF*, and *AD* medians of *ABC*.
(5) Given: *DH* equal and parallel to *BC* and cutting *AC*.
(6) To prove *HA* = and ∥ to *CF*.
(7) Sketching of a satisfactory figure (showing *EF, HE, AH*, and *HC*).
(8) *EBDH* is a parallelogram.
(9) If two sides of a quadrilateral are equal and ∥ the figure is a parallelogram.
(10) *EF* is ∥ and ½*CB* = *DB*.
(11) A line connecting the mid-points of two sides of a △ is ∥ to the third side and equal to ½ of it.
(12) *HE* = and ∥ to *DB*.
(13) Opposite sides of a parallelogram are equal and ∥.
(14) *FE* and *EH* are in the same straight line.
(15) But one straight line can be drawn through a given point ∥ to a given straight line.
(16) *HE* = *EF*. (Each equals ½*CB* = *DB*.)
(17) *CE* = *EA*. (Hypothesis.)
(18) *CHAF* is a parallelogram.
(19) If the diagonals of a quadrilateral bisect each other, the figure is a parallelogram.
(20) *CF* is = and ∥ to *HA*.
(21) The opposite sides of a parallelogram are = and ∥.
(22) The medians of a △ intersect at a common point (inferred from figure).
(23) Assumed *AFCH* a ▱, then said *HC* ∥ to *AF*.
(24) ∠ *HCA* = ∠ *CAF* (alt. int. ∠s) *HC*=*AF* (given), then
(25) △ *AHC* = △ *AFC* (S ∠ S) ∠ *ACF*=∠*HAC*. (Hom. parts.)
(26) Then, *HA* ∥ *CF* (alt. int. ∠s).
(27) Lines are ∥ if alt. int. ∠s are =.
(28) Random statements correct *per se*, but of no possible use.
(29) If lines are ∥ alt. int. ∠s are ∥, or corresponding ∠s are =.
(30) △ *HEA* = △ *CEF* (S ∠ S). Based on *FC*=*AH* and ∥.
(31) Assumed *FEH* a straight line.
(32) *ED* is = and ∥ to *FB*. Line between mid-points of two sides of △ is ∥ and = third side.

(*Continued on Pages 28 and 29*)

RECORD ON USE MADE OF FACTS

	Total		*Correct*		*Unsatisfactory*		*None*		*Wrong*	
	Boys	*Girls*	*Boys*	*Girls*	*Boys*	*Girls*	*Boys*	*Girls*	*Boys*	*Girls*
(1)	162	309	162	309	0	0	0	0	0	0
(2)	162	309	162	300	0	4	0	0	0	5
(3)	162	308	161	305	0	1	1	0	0	1
(4)	162	309	149	286	0	0	12	23	1	0
(5)	162	309	153	277	1	3	8	27	0	2
(6)	162	309	150	282	1	1	9	26	2	0
(7)	154	309	83	110	17	29	45	124	9	46
(8)	151	297	51	34	1	1	98	262	1	0
(9)	149	296	49	38	3	2	94	256	3	0
(10)	144	295	35	14	3	5	104	276	2	0
(11)	137	304	40	17	1	3	96	276	0	8
(12)	137	295	29	19	12	11	96	265	0	0
(13)	137	297	33	22	3	8	98	266	3	1
(14)	143	289	12	2	0	0	124	282	7	5
(15)	141	287	10	2	1	0	127	283	3	2
(16)	146	290	24	5	2	7	120	278	0	0
(17)	151	292	37	32	1	0	113	260	0	0
(18)	154	288	41	32	0	1	113	255	0	0
(19)	139	278	17	7	1	2	120	267	1	2
(20)	156	299	46	38	4	13	106	248	0	0
(21)	120	265	31	30	2	1	85	232	1	2
(22)	154	306	121	234	1	7	3	8	24	57
(23)	15	42	0	0	0	0	0	0	15	42
(24)	5	26	2	9	0	2	0	1	3	14
(25)	12	39	1	11	1	2	0	0	10	26
(26)	5	19	2	9	0	0	0	0	3	10
(27)	8	20	7	20	0	0	0	0	1	0
(28)	11	32	0	0	0	0	0	0	11	32
(29)	7	30	7	19	0	0	0	0	0	11
(30)	11	24	0	1	0	0	0	0	11	23
(31)	16	30	0	0	0	0	0	0	16	30
(32)	5	1	4	0	0	0	0	0	1	1

EXERCISE 13—(*continued*)

ABC is a triangle and AD, BE, and CF are its medians. DH is drawn equal and parallel to BE, and cuts AC. Prove that HA is equal and parallel to CF.

FACTS AND PRINCIPLES USED OR NECESSARY TO METHOD ADOPTED

(33) *BD* = *FE* (‖s between ‖s are =).

(34) Used 1–18 above; proved △ *AEF* = △ *HEC* (S ∠ S). Then *HC* = *AF*, ∠ *HCE* = ∠ *EAF*, *HA* ‖ *CF* (alt. int. ∠s).

(35) *HAFC* is a ▱ (opposite sides are = and ‖). *HA* = and ‖ *CF* (opposite sides of a ▱ are = and ‖). Then

(36) Drew *HC* = to *AF* and ‖ to it.

(37) Used 1–17, proved △ *AEH* = △ *FEC*, *AH* = *FC* and ∠ *AHE* = ∠ *EFC* (hom. parts) *HA* ‖ *CF* (alt. int. ∠s).

(38) Used 1–6, drew *ED* and *HC*, proved △ *BED* = △ *DHC*, ∴ *ED* = *HC*, *ED* = ½*AF*.

(39) Drew *ED* and *HC*, proved *HCDE* a ▱, *HC* = and ‖ *AF*.

(40) Opposite ∠s of ▱ are =.

(41) Proved *EAFD* a ▱ only.

(42) Proved *EFBD* a ▱ only.

RECORD ON USE MADE OF FACTS

	Total		*Correct*		*Unsatisfactory*		*None*		*Wrong*	
	Boys	*Girls*	*Boys*	*Girls*	*Boys*	*Girls*	*Boys*	*Girls*	*Boys*	*Girls*
(33)	4	6	2	3	2	1	0	0	0	2
(34)	2	2	2	0	0	1	0	0	0	1
(35)	3	0	2	0	1	0	0	0	0	0
(36)	2	3	0	0	0	0	0	0	2	3
(37)	1	4	1	1	0	0	0	0	0	3
(38)	2	1	1	1	1	0	0	0	0	0
(39)	15	4	15	4	0	0	0	0	0	0
(40)	3	1	2	1	1	0	0	0	0	0
(41)	1	0	0	0	1	0	0	0	0	0
(42)	4	0	3	0	1	0	0	0	0	0

CHAPTER II

THE RELATIVE DIFFICULTIES OF THE EXERCISES

Purpose of the Chapter. In this chapter we shall consider various measures of the relative difficulties presented to pupils by the several original exercises of the examination. The measures used here are admittedly crude measures of the relative difficulties of the exercises. If one of the exercises had presented just no difficulty to the pupils, that is, if all had solved it correctly, we could, on the assumption that abilities of the pupils were normally distributed, scale all of the exercises as to difficulty. It would then have been possible to state just how many times more difficult one exercise was than another for pupils of the tenth grade.

Since our data show that no one problem is solved correctly by all pupils, and since the scores (see page 106) made by pupils are not normally distributed, it is not possible therefore to express directly and precisely the relative difficulties of the exercises. We shall, however, express their relative difficulties roughly in terms of such crude measures as:

1. The percentages of pupils who select each of the several exercises as one of the eight units which they elect to be examined upon.
2. The mean and median scores attained by the entire group of pupils solving or attempting to solve each original exercise.
3. The percentage of pupils solving correctly each of the several exercises, and
4. A composite score which combines the mean scores made and the percentages of pupils selecting each of the several exercises.

Through the use of the several measures of difficulty stated above, but particularly through the composite score last mentioned, we shall show both by their estimation of relative difficulties and by actual achievement of success that the several exercises present to the pupils widely different degrees of difficulty. It will also be shown that Exercises 9 and 13 are respectively nine and fourteen times as difficult as Exercises 7 or 8.

It will then be pointed out that though measurable degrees of difficulty do exist, yet following the practice common to the grading of most examinations, the several exercises are given uniform credit. Although Exercise 13 is approximately fourteen times as difficult as Exercise 8, yet in scoring papers teachers and reviewers marked each exercise 12½ per cent, that is, of equal value, if correct.

Percentages of Selections. The study shows that it is the evident practice of both teachers and examiners to assign equal weights to the several exercises of an examination. Equal weights are assigned to exercises which are obviously unequal in difficulty, a practice as unjust as that of assigning unequal weights to exercises of equal difficulty. Here a crude measure of the relative difficulties of the exercise is found through a comparison of the percentages of pupils who selected each of the several exercises.

In taking the examination, it is not the purpose of the pupil to demonstrate his knowledge of the facts of geometry, but rather to gain the highest score possible under the rules set for the examination. Since his own record and to a less degree that of his school is at stake, the pupil feels it is his duty to select the exercises which he can do most easily and with the least chance of error. The percentages of pupils selecting the several exercises are then crude measures of the relative difficulties which pupils find in these exercises.

The following table shows the percentages of selections for the several exercises:

TABLE I

Exercise	No. 5	No. 6	No. 7	No. 8	No. 9	No. 10	No. 11	No. 12	No. 13
Papers Examined	1673	2197	1366	1243	2793	1776	1845	1848	2548
Times Selected	1153	1045	1124	1080	499	1016	1057	1117	471
Per Cent of Times Selected	69.9	47.6	82.3	86.9	17.9	57.2	57.3	60.4	18.5

Diagram I shows graphically the several degrees of difficulty recognized by pupils in these exercises as they make selections upon which to be examined. In increasing order of difficulty, as judged on this basis, the exercises stand as follows: Exercises 8, 7, 5, 12, 11, 10, 6, 13, and 9. From the standpoint of the preferences of the pupils, there is a wide range from Exercises 8 and 7 to Exercises 13 and 9. Although the order of presentation may have had some influence on the frequency of selection of the several exercises, this, as may be seen later, is largely overcome by the difficulties which are inherent in the exercises.

DIAGRAM 1. PERCENTAGES OF PUPILS SELECTING EACH EXERCISE

In order of preference as to difficulty the ranks are Exercises 8, 7, 5, 12, 11, 10, 6, 13, and 9. There seems to have been slight ground for choice between Exercises 10 and 11 and between Exercises 9 and 13.

Measures of Abilities Used in the Study. In writing the examination set by the New York Regents, pupils were instructed to select and to solve eight exercises from among the thirteen offered. Four of the thirteen exercises were theorems which are commonly proved in full in text books on plane geometry. The work done by pupils in reproducing these theorems was not examined in this study. The work done by each pupil in solving the four or more originals which he chose to solve, was examined and the grades assigned by the teachers and approved by the reviewers were recorded. To the correct solution of each exercise, a grade of 12½ per cent had been assigned, while the grade zero had been given in cases where practically nothing was done on the solution of an exercise even where this exercise was one of the eight exercises selected.

On the basis afforded by the grades assigned by teachers, measures of the abilities shown by boys and girls are computed. The distribution of scores obtained in this manner is reported on page 106 and shown graphically for Exercise 10 in Diagrams 9 and 10 (see page 49). It is a fairly common practice to use the median score attained by a group as a measure of the ability of the group in the trait in question. This measure is approved as a measure of central tendency where the scores made are normally distributed. The median score is easy of computation and where the scores show that a group of sufficient size has been measured to afford a random sampling of children in the trait in question, the use is to be commended.

However, the frequencies, as shown on page 106 and as graphed for Exercise 10, show distributions far from the normal, in fact they are badly skewed. It should be recalled that the group of papers examined in no case includes papers of pupils who had failed in the examination as a whole. Records made show that papers are usually graded wrong, half right, or correct. It should be recalled also that in selecting exercises for solution the pupils had the privilege of rejecting five original exercises of nine given. We do not, therefore, expect to find a normal distribution of scores, by which we mean a relatively small number of low scores, more poor scores, a relatively large number of medium scores, a smaller number of superior scores, and a still smaller number of excellent scores.

As a measure of central tendency, the median score indicates the point below and above which 50 per cent of the cases lie. Again the computation of the median (see page 105) does not take into account the relative values of the scores made. A measure more suitable as a measure of central tendency for a skewed distribution and one which does take into account the size of each score made is the *mean* or average. This measure, though more tedious to compute (see page 105) has, because of its worth here, been adopted as the measure of central tendency in each distribution of the study. For purposes of comparison, however, both mean and median scores have been computed and will be given. The probable errors of these measures have been computed and appear in connection with the measures on page 113.

A table showing the measures of central tendency, variability, and correlation with the probable error of each measure appended is given on page 113. The number of cases used in the study (approximately 1000 in each exercise, save Exercises 9 and 13) is sufficient to give a satisfactory degree of reliability to each measure used, as is seen by a comparison of such measures with their appended probable errors. The relative size of each measure of central tendency, variability, and correlation, when considered in connection with its probable error, shows the degree of reliance which may be placed upon the measure as such.

Median and Mean Scores Attained. As a second measure of difficulty, the median scores might be taken. However, the distributions for the several problems are badly skewed (see page 105). Where distributions are far from that shown by the normal surface of frequency, the mean is a better measure of the central tendency than the median (see page 33). In using median scores the weighted scores made by individuals are not considered. In the mean score, however, the size of the score attained by each pupil is considered.

The following table shows the relative difficulties of the exercises as measured in terms of median and mean scores attained by pupils. (The range in possible scores for each exercise is 0—12.5.)

TABLE II

Exercise	No. 5	No. 6	No. 7	No. 8	No. 9	No. 10	No. 11	No. 12	No. 13
Median	12.26	10.07	11.92	12.29	5.88	9.01	8.88	12.26	2.87
Mean	9.88	9.06	9.77	9.74	5.43	9.11	7.80	10.58	3.20

The data of Table II are presented graphically in Diagram 2. From both table and diagram the wide range in difficulties from

Exercises 5, 8 and 12 to Exercises 9 and 13 shows notable differences in difficulty.

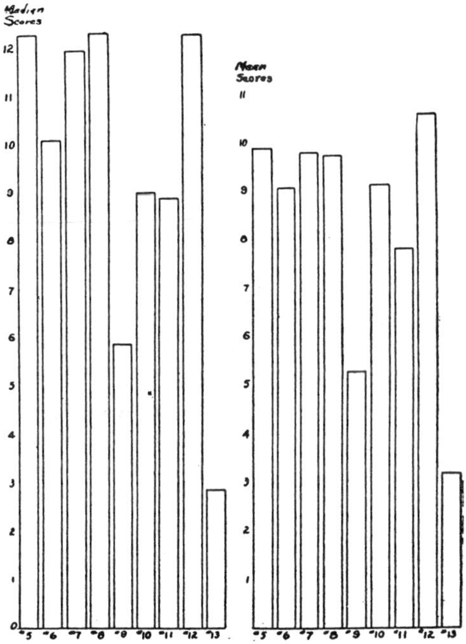

DIAGRAM 2

MEDIAN SCORES

High scores are made on Exercises 5, 8, 7, 12. Note the difference between these scores and those for Exercises 9 and 13.

MEAN SCORES

Note the high mean scores of Exercises 5, 6, 7, 8, 10, 12, as contrasted with low scores for Exercises 9 and 13.

Table II and Diagram 2 give the relative ranks in the difficulties of the exercises as shown in the median and mean scores attained by the pupils. Using the highest median and mean scores as indicative of the least difficulty, the table below shows (in ranks of difficulty) the relative difficulties of the several exercises.

Table III shows clearly that, on the whole, there is close agreement between the rankings of the exercises obtained through the use of the median and mean scores as measures of difficulty of these exercises.

TABLE III

Exercise as Ranked by	No. 5	No. 6	No. 7	No. 8	No. 9	No. 10	No. 11	No. 12	No. 13
Median Scores	2.5	5	4	1	8	6	7	2.5	9
Mean Scores	2	5	3	4	8	6	7	1	9

Percentage of Exercises Correctly Solved as a Measure of Difficulty. In an examination such as this where pupils have an option in choice of exercises, the difficulty to all pupils can not be measured directly. We have already considered the degree of avoidance, the median and mean scores made as measures of difficulty and will discuss here just one more feature of the results, the percentage of correctly solved exercises.

The following table shows these data for each of the several exercises.

TABLE IV

Exercise	No. 5	No. 6	No. 7	No. 8	No. 9	No. 10	No. 11	No. 12	No. 13
Percentage of Correct Solutions	57.5	39.4	54.5	64.8	1.8	36.4	8.2	68.4	6.6

That exercises are correctly solved, or practically so, by widely different percentages of those who select them indicates a wide difference in the real difficulties inherent in the exercises. When one exercise is correctly solved by 68.4 per cent and another by 1.8 per cent, the evidence is conclusive that marked differences in difficulty do exist. This fact in itself is not important, for differences must of necessity be present, but the common practice of assigning equal weights in rating all questions is not even a questionable policy. It contradicts the facts in the case and is therefore obviously wrong.

Times Values in Relative Difficulties. Up to this point we have considered (1) the percentages of pupils selecting an exercise, (2) the median and mean scores made by pupils in solving exercises, and (3) the percentages of pupils making perfect scores, as measures of the difficulties presented to the pupils by the several exercises. We shall now combine certain measures in order to show approximately just how many times more difficult one problem is than another. Since the mean score and the percentage making a perfect score measure approximately the same function, we shall not consider the percentages making perfect scores in our composite score upon which we shall base relative difficulties.

Now as the percentage of selections (P_s) increases, the difficulty (D) of the exercise decreases. We may therefore say that D varies inversely with P_s, or

$$D \text{ varies with } 1/P_s.$$

Again, as the mean score (M) increases, the difficulty (D) of the exercise decreases. We may say therefore that D varies inversely with M, or

$$D \text{ varies with } 1/M.$$

Since D varies with $1/P_s$ and with $1/M$ we may say that D varies jointly with $1/P_s \times 1/M$, or

$$D = K \times 1/P_s \times 1/M = K \times 1/(P_s \times M).$$

If it were possible to estimate the influence of all other factors (such as the position of the exercise on the list, the frequency of the use made of each fact involved, etc.), which enter into the determination of the difficulty of an exercise, we should then have a value for our constant K and should be able to compute more exactly the relative difficulties of the several exercises. We have in the above discussion considered two essential features, and the best assumption we can make under the circumstances is that other factors so counteract one another that their combined influence would not greatly disturb the results obtained from the use of the mean scores and the percentages of the pupils selecting the exercise. On this hypothesis K equals 1, which gives the following measure of difficulty for the several exercises:

$$D = 1/(P_s \times M).$$

This composite value of D for each of the several exercises is computed for Exercise 5 as follows:

P_s or Percentages of Selections = 68.9 per cent and
M or Mean Score Attained = 9.88 per cent

$$\text{Then } D = \frac{1}{P_s \times M} = \frac{1}{0.689 \times 0.0988} = 14.7 \text{ units}$$

The following table shows the composite measures of difficulty for the several exercises.

TABLE V

Exercise	No. 5	No. 6	No. 7	No. 8	No. 9	No. 10	No. 11	No. 12	No. 13
Difficulty Units	15	23	12	12	103	19	22	16	169

The above values for D are represented graphically in Diagram 3.

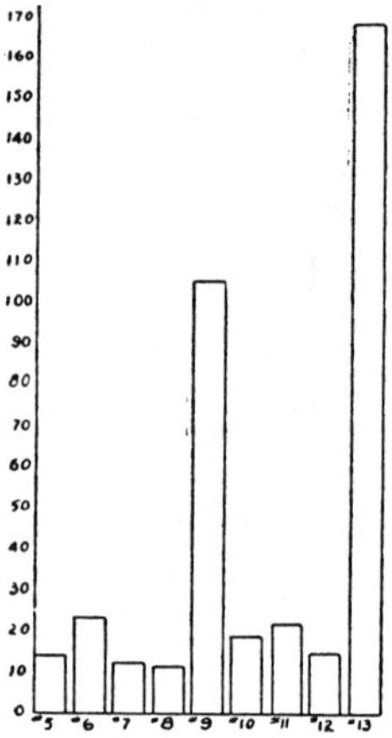

DIAGRAM 3. RELATIVE DIFFICULTY IN TERMS OF MEAN SCORES AND PERCENTAGES OF SELECTION. $D = \frac{1}{M} \times \frac{1}{P_s}$.

From the above diagram it appears that Exercise 13 is about fourteen times as difficult as Exercise 8, etc.

From Diagram 3 it appears that Exercises 5, 7, 8 and 12 present approximately equal difficulty to pupils and Exercises 6, 10, and 11 seem of approximately equal difficulty. Exercise

9 appears to be approximately seven times as difficult as any exercise in the first group mentioned above, while Exercise 13 is approximately eleven times as difficult as any exercise in the first group and approximately fourteen times as difficult as Exercise 7 or Exercise 8.

As has been said above, to assign equal credit to the correct solution of exercises as unequal in difficulty as Exercises 7, 9, and 13 is to equate difficulties which are far from equal in value. This practice should be abandoned by all teachers of geometry.

Conclusions. A suggestion to teachers and examiners comes out of the above data. It is this: In setting up a test or examination where pupils are to be rated and ranked, or classes or schools compared, questions should be weighted equally or assigned equal credit only when they are almost certain to present approximately equal difficulties to the majority of the pupils. Where options are offered, selections should be allowed from among problems of approximately equal difficulty.

A critical examination of the several steps involved in the solutions of several exercises will provide for the teacher data upon which to base judgments as to the approximate relative difficulties. It is therefore incumbent upon those who set examinations to write down in detail each step of each solution requested and to estimate the difficulties involved. Such a procedure should prevent such gross errors in assigning proper credit to work done as appear in the case under consideration in this study.

CHAPTER III

SEX DIFFERENCES IN GEOMETRIC ABILITIES CONTRASTED AND CORRELATED

Purpose of the Chapter. It is the purpose of this chapter to point out sex differences in interests in selecting and in abilities to solve correctly certain original exercises in geometry and then to express these differences by means of coefficients of correlation. It will be our purpose to point out how, if at all, school practice should be altered in the light of this and related studies.

We shall first measure the sex differences found in the selections of exercises for solution made by the sex groups; second, we shall point out the relative difficulties experienced by the groups in solving the several exercises; third, we shall note the sex differences shown by the percentages of each group making perfect scores on each of the several exercises; fourth, we shall express differences in difficulties experienced by boys and girls in times values; and fifth, we shall measure the sex differences in abilities to solve originals by showing the percentage of one group which equals or exceeds the ability of the median member of the other group.

We shall then relate the preferences shown by the sexes for certain exercises by means of a coefficient of correlation—the tetrachoric *r*. Then we shall relate the abilities shown by the sexes in solving originals by means of still another coefficient of correlation—the bi-serial *r*.

As has been stated previously, it was our aim to examine enough papers to provide approximately five hundred cases of each exercise for each sex. The papers at hand (approximately 2800 papers of pupils who passed the examination as a whole) did not always provide the desired number of cases for each exercise; yet on the basis of papers available and examined, we are able to measure and relate the sex differences both as to preference for certain types of exercises and as to ability to solve the exercises selected.

Briefly, the findings of the chapter cover the following items:

1. The least difference found between the percentages of boys and girls selecting an exercise is in Exercise 5, a numerical exercise and the first exercise in the list. The greatest difference in percentage of selection is in Exercise 8, one involving construction. This type of exercise seems to make a stronger appeal to boys than to girls.

2. Boys experience somewhat less difficulty in solving originals than do girls, as is seen through the use of each of several measures employed,—the mean and median scores, the percentages of perfect scores attained, the composite score, and the percentages of boys who equal or excel in ability the median girl.

3. The sex differences in abilities to solve originals are all small, in fact so small and so uncertain as to give no ground for segregating the sexes, or for assigning to one sex group one type of exercise and to the other group another type of exercise.

4. The probability of a sex difference existing in the selection of certain of the given exercises varies from 1:1 to 1:559,000 or from a probability that no sex difference exists in the choice of Exercise 5 to a practical certainty that a sex difference does exist in Exercise 8, the construction exercise.

5. The correlations (tetrachoric *r*'s) between being a boy (rather than a girl) and selecting certain exercises (rather than rejecting them) vary from + 0.315 to — 0.128, showing thereby that boys prefer the construction type of exercise and avoid most decidedly Exercise 13 which is an original involving proof. This last exercise has been previously shown to be approximately fourteen times as difficult as each of several other exercises in the list. The positive correlation of + 0.35 probably indicates a real sex difference in the preference boys have for construction exercise, and the negative correlation of — 0.13 may probably indicate a sex difference in clearness of recognition of difficulties inherent in a problem as in the case of Exercise 13.

Sex Differences Contrasted

Percentage of Each Sex Selecting Each Exercise. Table VI below gives the percentages of boys and girls selecting each exercise, and the data upon which these percentages are computed.

TABLE VI

PERCENTAGES OF BOYS AND GIRLS SELECTING THE SEVERAL EXERCISES

Exercise	Papers Examined. Boys			Papers Examined. Girls		
	Total No.	Times Selected	Per cent of Times Selected	Total No.	Times Selected	Per cent of Times Selected
No. 5	792	546	68.9	881	607	68.9
No. 6	1030	518	50.3	1167	527	45.2
No. 7	621	498	80.2	745	626	84.0
No. 8	540	501	92.8	703	579	82.4
No. 9	1289	256	19.8	1504	243	16.2
No. 10	842	464	55.1	934	552	59.1
No. 11	924	506	54.8	921	551	59.8
No. 12	827	547	66.1	1021	570	55.8
No. 13	1063	162	15.2	1485	309	20.8

The above data on the selection and rejection of the several exercises are shown graphically in Diagram 4 on page 43.

From Table VI and Diagram 4, it appears that on the basis of percentages of times selected, boys and girls agree as to the difficulty of Exercise 5.

The largest differences in selections appear in Exercise 8 and Exercise 12, in which cases boys show a decided preference for each of these exercises. The differences for the other exercises are not so clearly marked. Yet it appears that boys judge Exercises 6, 8, 9, and 12 less difficult than do girls.

The following table (based on the percentages given above) shows the order of difficulties seen in the exercises by boys and girls. (Rank 1 means greatest in scale of difficulty).

TABLE VII

Exercise	No. 5	No. 6	No. 7	No. 8	No. 9	No. 10	No. 11	No. 12	No. 13
Ranked by Boys	7	3	8	9	2	5	4	6	1
Ranked by Girls	7	3	9	8	1	5	6	4	2

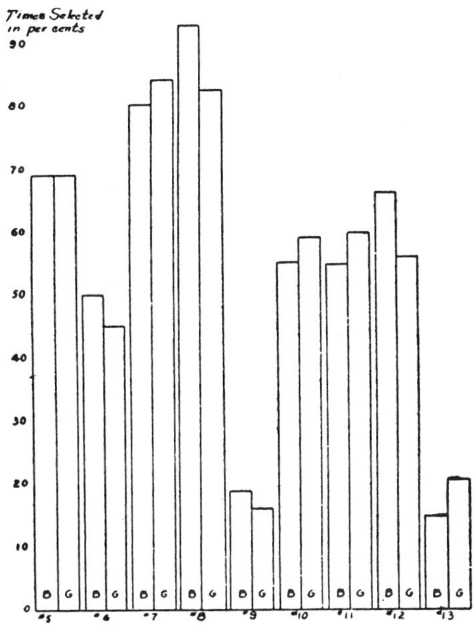

DIAGRAM 4. RELATIVE DIFFICULTIES OF EXERCISES FOR BOYS AND FOR GIRLS AS SEEN IN THE DIFFERENCES OF THE PER CENTS OF TIMES SELECTED

Boys and girls agree in their judgments as to the difficulty of Exercise 5. Boys judge Exercises 6, 8, 9, and 12 less difficult than do girls.

There is a relatively close agreement between boys and girls as to their placements of the several exercises in order of difficulty. Some of the significant points as seen in Table VI and Table VII are:

1. There is an agreement in the percentages of the sexes selecting Exercise 5.

2. The greatest difference is that between percentages of boys and girls in selecting Exercises 8 and 12.

3. In three out of the nine exercises boys and girls agree as to order of difficulty which they see in the exercise.

4. In four out of the nine exercises there is a displacement of but one step and,

5. In but two out of the nine exercises there is a displacement of two steps.

The degree cf relationship existing between boys and girls in their selection of exercises which they elect to be examined upon is expressed on page 53 by tetrachoric r's or fourfold correlations. The r of this type is to be interpreted as is the Pearson product-moment coefficient of correlation between two groups measured in the same trait.

Median and Mean Scores Attained. The mean scores attained by boys and by girls on each of the several problems are computed as on page 105. The median scores are also given there and are repeated in Table VIII.

TABLE VIII

Exercise	No. 5	No. 6	No. 7	No. 8	No. 9	No. 10	No. 11	No. 12	No. 13
Boys' Median Score	12.29	10.21	11.99	12.32	6.01	10.75	9.02	12.27	2.97
Girls' Median Score	12.10	9.92	11.21	12.26	5.58	7.48	8.80	12.23	2.83
Boys' Mean Score	10.08	9.36	10.20	10.68	5.85	9.68	8.33	11.02	4.29
Girls' Mean Score	9.70	8.76	9.42	8.92	4.98	8.62	7.34	10.16	2.63

The above median and mean scores for the sex groups are shown in Diagrams 5 and 6 on page 45.

From Diagram 5, presenting results in terms of the median scores, it appears that boys as a group experience less difficulty in solving original exercises than do girls. Diagram 6, presenting the mean scores, brings out the same fact and shows in addition that the differences in ability between boys and girls are greatest in those exercises (the construction exercises excepted) which both groups judge very difficult, as shown by the small percentages of selections for these exercises.

Again, the least difference in mean scores appears in that exercise which the groups agree upon as the easiest exercise.

Percentages of Exercises Correctly Solved as a Measure of Difficulty to the Sexes. Table IX is based on the distributions of page 106 and shows the difficulties of the several exercises for boys and girls in terms of the percentages of each group attaining perfect scores, that is 12 per cent or 12½ per cent.

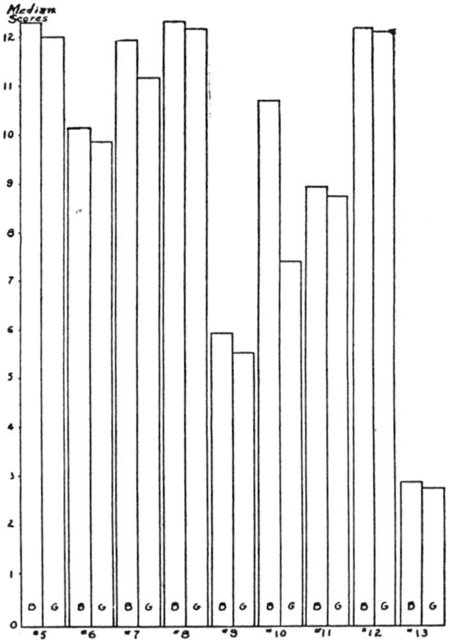

DIAGRAM 5. MEDIAN SCORES ATTAINED BY BOYS AND GIRLS

Boys made higher median scores in each exercise than did girls.

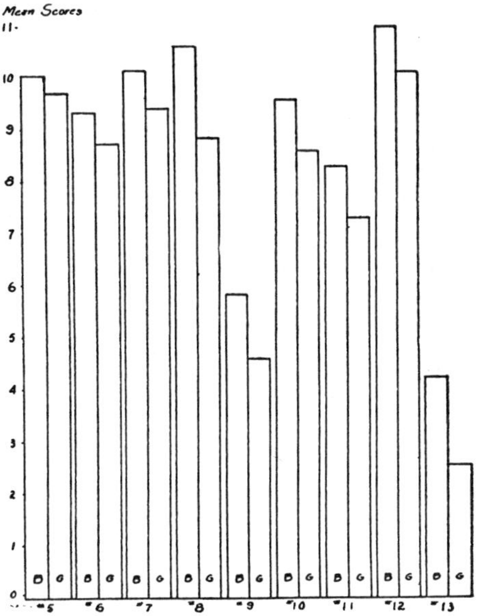

DIAGRAM 6. MEAN SCORES ATTAINED BY BOYS AND GIRLS

The diagram shows that boys made higher scores in each exercise than did girls. Note that these differences are greater than in median scores.

TABLE IX

Exercise	No. 5	No. 6	No. 7	No. 8	No. 9	No. 10	No. 11	No. 12	No. 13
Per Cent of Boys Making Perfect Scores	61.5	44.6	61.2	76.4	1.95	45.5	12.3	72.8	14.8
Per Cent of Girls Making Perfect Scores	53.9	34.3	49.2	54.7	1.65	28.8	4.54	64.2	2.27

The data of the above table are shown in Diagram 7.

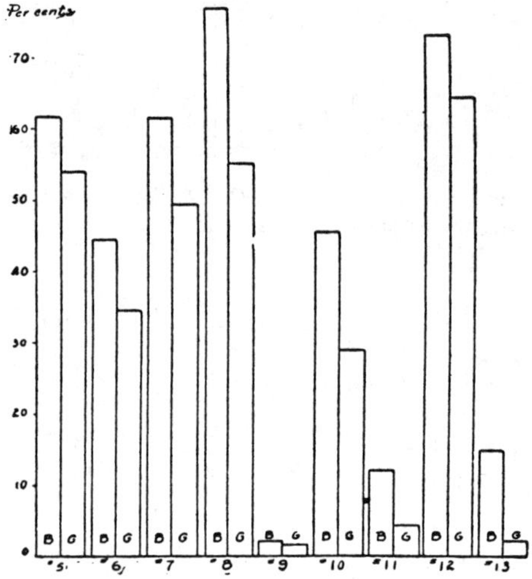

DIAGRAM 7. PER CENTS OF BOYS AND GIRLS MAKING PERFECT SCORES

Note that a larger percentage of boys made perfect scores on each exercise than did girls.

From this table showing the percentage of perfect scores in the several exercises as made by boys and girls and from Diagram 7, it appears that both boys and girls find greater difficulty in attaining prefect scores in Exercises 9, 11 and 13 than in Exercises 5, 6, 7, 8, 10, and 12. Each group finds Exercise 8 and Exercise 12 far easier than either Exercise 9 or 13. The fact that in each problem a higher per cent of boys than of girls attains the perfect scores gives some additional evidence on the superior geometrical ability of boys.

Times Values of Differences in Difficulties for Boys and for Girls. The data showing per cents of boys and of girls selecting each exercise are combined with data showing mean scores made by each group into a composite measure for each sex group on each exercise, as was done for the combined group on page 37.

Upon these composite measures we shall base our final determinations of measures of difficulties presented to boys and to girls by each exercise. These composite measures (computed as on page 37) are given in Table X, and shown graphically in Diagram 8.

TABLE X

Exercise	No. 5	No. 6	No. 7	No. 8	No. 9	No. 10	No. 11	No. 12	No. 13
Difficulty Units (Boys)	14.4	21.2	12.2	10.1	86.2	18.8	21.9	13.7	153.4
Difficulty Units (Girls)	15.0	25.3	12.6	13.6	123.5	21.0	22.8	17.6	182.8

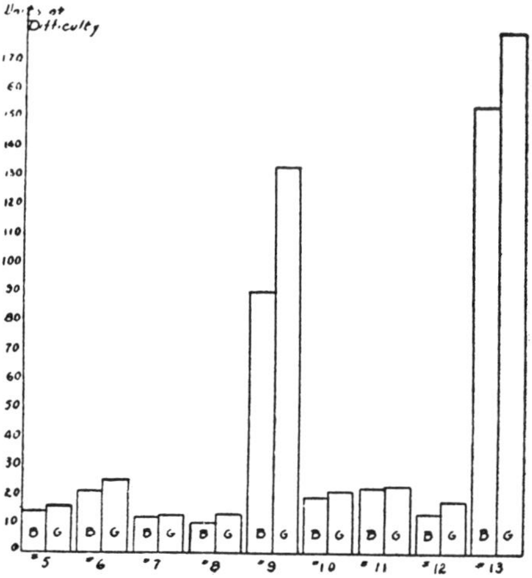

DIAGRAM 8. RELATIVE DIFFICULTY IN TERMS OF MEAN SCORES AND PERCENTAGES OF TIMES SELECTED

Each sex finds Exercises 9 and 13 much more difficult than the other exercises. Note again that boys find less difficulty in each exercise than do girls.

From the above table and diagram, the following facts appear:

1. Each exercise presents less difficulty to boys than to girls.

2. The more difficult exercises present greater differences in difficulty to the two sexes.

3. On the whole, the sexes agree as to the relative difficulties of the exercises.

4. The sexes agree that Exercises 9 and 13 are much more difficult than are the other exercises in the list.

5. In no cases save Exercises 9 and 13 are relative difficulties presented greater than 4 : 3 approximately.

Variability of the Sexes. In the Appendix, proof is given of the fact that the sigma (Standard Deviation) of a distribution provides a more reliable measure of variability of the trait in question than does the quartile deviation. The sigmas of the several distributions are computed on page 105 and tabulated on page 113. In order that comparable measures may be had, the coefficient of variability of each sex group (sigma divided by the mean) has been computed for each exercise. Then, finally, the ratio of boy-to-girl variability on each exercise has been found. These ratios for the several exercises are here given.

TABLE XI

Exercise	No. 5	No. 6	No. 7	No. 8	No. 9	No. 10	No. 11	No. 12	No. 13
Ratio of Boy-to-Girl Variability	0.95	0.88	0.79	0.67	0.92	0.89	0.88	0.72	1.09

From Table XI it appears that in solving eight of the nine exercises, boys are less variable than girls. Combining the above, the data show that for the groups considered and for the trait in question, boys are 0.87 as variable as girls. This finding does not support the finding reported by Thorndike (*Educational Psychology,* Vol. III, page 194) ". . . in Regents' Examinations in Latin, English, and in History, it" [girl-to-boy variability] "was 0.96; in school marks in eight subjects it was 0.86." If the difference in findings is to be accounted for, a possible explanation may lie in the fact that in this study the papers of both boys and girls who failed in the examination were not considered. Because of the large number of individuals here considered, the reliability of the findings is high for the group under consideration. The study shows clearly that among pupils who pass the Regents' Examination in plane geometry, boys are somewhat less variable than are girls in ability to solve geometric originals. As might be expected, the lowest ratio of boy-to-girl variability in achievement appears on the construction exercise, for which the greatest difference in preference was shown.

The distribution of scores on Exercise 10 are typical of the distributions of the study. These distributions are presented in Diagrams 9 and 10.

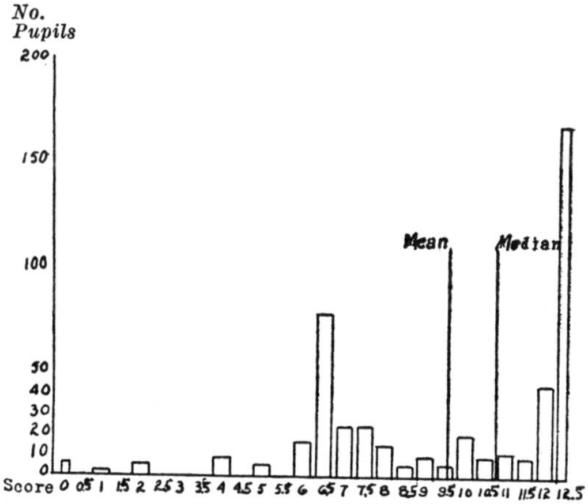

DIAGRAM 9. SCORES OF BOYS. EXERCISE 10

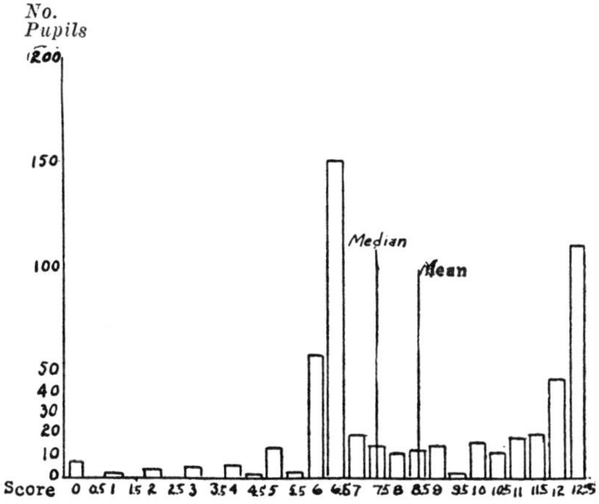

DIAGRAM 10. SCORES OF GIRLS. EXERCISE 10

The foregoing diagrams show that the distributions for each sex are badly skewed. Again, it can be seen that in the boys' diagram there is a greater tendency for scores to pile up at the upper end of the scale than in the girls'. This means of course a lower male variability.

Table XI shows almost as wide divergence (0.79 to 0.95) in boy-to-girl variability on numerical exercises as on exercises involving demonstration (0.72 to 1.09). The data show, therefore, that one of these types is no better adapted to the abilities of one sex than to the other. The foregoing supports the conclusion of page 51.

Sex Differences Expressed in Terms of the Percentage of One Group Which Equals or Exceeds the Median of Another. From the distributions on page 106, the data of Table XII have been deduced. The table indicates for each exercise of the list the percentage of boys who equal or exceed in ability the median girl.

TABLE XII

Exercise	No. 5	No. 6	No. 7	No. 8	No. 9	No. 10	No. 11	No. 12	No. 13
Per Cent of Boys who Equal or Excel the Median Girl	57.3	56.2	62.2	66.3	64.1	67.0	60.3	56.3	59.3

With due regard to the fact that the percentages reported abòve do not include records of pupils who failed in the examination as a whole, we conclude that in ability to solve original exercises in geometry the median boy surpasses the median girl.

While the results reported in this study show greater sex differences than those reported in previous investigations, yet the differences between pupils in each group as observed from the distributions of page 105 show that individual differences in each sex group far outweigh in importance the differences between the sexes.

On the above point Thorndike (*Educational Psychology,* Volume III, page 184) says: "The most important characteristic of these differences (differences shown from investigations on scholarship in relation to sex) is their small amount. The individual differences within one sex so enormously outweigh the differences between the sexes in these intellectual and semi-intellectual traits that for practical purposes the sex difference may be disregarded. So far as ability goes, there could hardly be a stupider way to get two groups alike within each group but differing between groups than to take the two sexes. As is well known, the experiments of the past generation in educating women have shown their equal competence in school work of elementary, secondary school, and collegiate grade.''

Stockard and Bell in a study of abilities of high school and normal school pupils in geometry reported through the *Journal of Educational Psychology*, Vol. VII, as follows: ''The preliminary test seemed to point to the conclusion that on finishing plane geometry girls made the higher scores.'' In the report on the same study these authors assert that their results indicate that ''in geometrical abilities boys and girls are about equal and that in general boys are slightly in the ascendancy.'' The scores (boys 54.5 per cent correct, and girls 64.0 per cent correct) here quoted were based on the percentages of boys and girls who solved their geometry test exercises correctly.

The second finding quoted above is supported by that of this study, while the first finding is not supported by the data of this study. The number of pupils considered in the Stockard and Bell study (372) does not give to their results the degree of reliability which this study affords, because of the larger number of cases here considered.

Miss Rusk,[1] in a study of grades received for scholarship in high-school studies, found that in geometry 53 per cent of the boys reached or exceeded the median girl.

Thorndike,[2] in a study of abilities of boys and girls in the same classes in high schools, found that in mathematics (Regents' examinations and school marks) 57 per cent of the boys reached or exceeded the median ability for girls.

In commenting on the results of the last two studies quoted above, Thorndike finds no discernible difference between boys and girls, in the case of mathematics. This study, however, does show that the median boy slightly excels the median girl in ability to solve each of nine original exercises in plane geometry.

By referring to Chapter I, page 3, it will be noted that Exercises 5, 6, 7, and 9 are of a numerical character, that Exercises

1 Thorndike, E. L., *Educational Psychology,* Vol. III, p. 184.

2 *Ibid.*

10 to 13 inclusive require proof involving a syllogistic organization of facts, and that Exercise 8 is one which involves construction. By comparing the percentages of boys who equal or exceed in ability the median girl, it appears that the range in percentages of boys who equal or exceed in ability the median girl on numerical exercises (56.2 per cent to 66.3 per cent) is not materially different from that found (56.3 per cent to 67 per cent) in exercises involving proof. The average in the former case is 60.0 per cent and in the latter 60.8 per cent. It follows then that in solving numerical exercises boys show no greater superiority over girls than in solving exercises involving proof and syllogistic organization. From this it appears that a proposal to offer numerical exercise work to one sex exclusively and proof exercises to the other would not be supported by data on the difference in abilities here observed.

Sex Differences Related

Probability of a Sex Difference in Choice of Exercises. By means of a method given by Pearson (*Phil. Trans.* Vol. 200 A. P. 25) and through the use of tables calculated by W. Palin Elderton (*Biometrika,* Vol. I, pp. 155–163) we are able to measure the probability of association between being a boy rather than a girl and selecting a certain problem rather than rejecting it. In this process we use the observed data for selections and rejections for each sex on each exercise and compare these data with the number which would have occurred on the basis of pure chance. This number is computed as is shown in the following fourfold table, where a, b, c, and d are observed values, and where a', b', c', and d' are such values as would have been expected on the basis of pure chance.

TABLE XIII

	Boys	Girls	Total
Selections	$a\ (a')$	$b\ (b')$	$a+b$
Rejections	$c\ (c')$	$d\ (d')$	$c+d$
Total	$a+c$	$b+d$	N

In the above table,

$$a' = (a+b) \times (a+c) \div N, \quad c' = (c+d) \times (a+c) \div N$$
$$b' = (a+b) \times (b+d) \div N, \quad d' = (c+d) \times (b+d) \div N$$

Through the use of the differences which are found between observed and chance values, Pearson has suggested the method of finding the probability of association which is also used in our computations.

Correlations in Choices of Exercises Expressed in Terms of the Tetrachoric r. Because of the desirability of expressing the degree of relationship by means of a correlation coefficient of known reliability, we have made use of the tables by Everitt (*Biometrika,* Vol. VI, pages 437-451), and computed these tetrachoric r's. Through these coefficients, we are able to express the relationship between being a boy rather than a girl and the selection of the exercise rather than the rejection of it. This method of work is explained in detail in connection with the computation of the r of Exercise 5 which follows:

Exercise 5

To compute the probability that being a boy rather than a girl leads to selecting the exercise.

Exercise 5	Girls	Boys	Totals
Times Selected	a 607 a'(607)	b 546 b'(546)	1153
Times Rejected	c 274 c'(274)	d 246 d'(246)	520
Totals	881	792	1673

Numbers in parentheses above are those which would occur on the basis of pure chance. $a =$ observed selections, $a' = (1153)(881) \div 1673$; etc.

$$X^2 = \frac{(607 - 607)^2}{607} + \frac{(546 - 546)^2}{546} + \frac{(274 - 274)^2}{274} + \frac{(246 - 246)^2}{246}.$$

$X^2 = 0.$

Then (see Pearson's tables for transmuting from X^2 to P):

$P = 1.$

Then the probability that a random sampling would lead to as large or larger deviation between theoretical (chance) and observed values (times selected) is 1 : 1.

Solving for r_t, the tetrachoric r or the coefficient of fourfold relationship:

Here

$$\frac{d}{n}=\frac{246}{1673}=0.14704;\ \tfrac{1}{2}(1-a_1)=T_0=\frac{792}{1673}=0.4734;\ \tfrac{1}{2}(1-a_2)=T_0'=\frac{520}{1673}=0.3108.$$

Then (formula and tables by Everitt):

For $T_0=0.4734$, $T_1=0.3980$, $T_2=0.0190$, $T_3=-0.1617$, $T_4=-0.0164$, $T_5=0.1080$, $T_6=0.0150$.

For $T_0'=0.3108$, $T_1'=0.3532$, $T_2'=0.1231$, $T_3'=-0.1091$, $T_4'=-0.0980$, $T_5'=0.0516$, $T_6'=0.0819$.

Then for

$$\frac{d}{n}=T_0T_0'+T_1T_1'r+T_2T_2'r^2+T_3T_3'r^3+T_4T_4'r^4+T_5T_5'r_5+T_6T_6'r_6,\ \text{etc.}$$

We have

$$0.14704=(0.4734)(0.3108)+(0.3980)(0.3532)r+(0.0190)(0.1231)r^2+(-0.1617)(-0.1091)r^3+(-0.0164)(-0.0980)r^4+(0.1080)(0.0516)r^5+(0.0150)(0.0819)r^6,\ \text{etc.}$$

Arranging terms and dividing by the coefficient of r,

$$0.00875r^6+0.00396r^5+0.01145r^4+0.12548r^3+0.01665r^2+r+0.00066=0.$$

It is evident that r is zero, for the more refined our computations, the more nearly zero does r become.

Now

$$\text{P.E.}_{r_t}=X_1\cdot X_{r_t}\cdot X_{a_1}\cdot X_{a_2}.$$

Here

$$X_1=\frac{0.67449}{\sqrt{1673}};\ \tfrac{1}{2}(1+a_1)=\frac{1153}{1673}=0.6892;\ \tfrac{1}{2}(1+a_2)=\frac{881}{1673}=0.5266.$$

$$X_1=\frac{0.67449}{40.9023},\ X_{r_t}=1.000;\ \text{for } \tfrac{1}{2}(1+a_1)=0.6892,\ X_{a_1}=1.3103;\ \text{for } \tfrac{1}{2}(1+a_2)=0.5266,\ X_{a_2}=1.2544.$$

Then

$$\text{P.E.}_{r_t}=\frac{0.67449}{40.9023}(1.000)(1.3103)(1.2544)=0.0271.$$

Then

$$r_t=0.00\pm 0.03.$$

On the basis of the data used, there appears to be just no correlation between being a boy rather than a girl in selecting this exercise. Such selections as were made might conceivably have been made by mixed groups.

Exercise 6

To compute the probability that being a boy rather than a girl leads to selecting the Exercise 6.

Exercise 6	Girls	Boys	Total
Times Rejected	640 (612)	512 (540)	1152
Times Selected	527 (555)	518 (490)	1045
Total	1167	1030	2197

$$X^2 = \frac{(-28)^2}{612} + \frac{28^2}{540} + \frac{28^2}{555} + \frac{(-28)^2}{490}$$

$$= 5.7455$$

Then

$$P = 0.1567.$$

The probability that a random sampling would lead to as large or larger deviation between theoretical (chance) and observed values (times selected) is 1 : 6.4.

Now r_t is found as follows:

$$\frac{d}{n} = 0.23578, \quad T_0 = 0.4688, \quad T_0' = 0.4756.$$

Then from the formula and tables by Everitt we have:

$$0.00145r^6 + 0.07350r^5 + 0.00177r^4 + 0.16488r^3 + 0.00234r^2 + r - 0.07830 = 0.$$

Here r is approximately 0.0783.

For this value of r, terms in powers of r higher than the third may be neglected.

Testing 0.0780 for r, we have

$$f(0.0780) = -0.00021.$$

Testing 0.0790 for r, we have

$$f(0.0790) = +0.00067.$$

Interpolating,

$$r_t = 0.0783.$$

$$\text{P.E.}_{rt} = X_1 \cdot X_{rt} \cdot X_{a1} \cdot X_{a2}.$$

$$= 0.0226.$$

Then

$$r_t = 0.08 \pm 0.02.$$

Therefore between being a boy rather than a girl and selecting this exercise rather than rejecting it, the coefficient of correlation (tetrachoric r) is 0.08 ± 0.02.

Summary on Sex Differences in the Choice of Certain Exercises. From solutions similar to the foregoing we have the following results in terms of probability of selections and of coefficients of correlation.

TABLE XIV

Exercise	*The probability that a random sampling would lead to as large or larger deviation between theoretical (chance) and observed values (times selected) is:	*The tetrachoric r or coefficient of correlation between being a boy rather than a girl and selecting the exercise rather than rejecting it.
No. 5	1: 1	$r_t = 0.00 \pm 0.03$ (no correlation)
No. 6	1: 6.4	$r_t = 0.08 \pm 0.02$
No. 7	1: 3	$r_t = -0.09 \pm 0.03$
No. 8	1: 559,000	$r_t = 0.32 \pm 0.04$
No. 9	1: 9.4	$r_t = -0.09 \pm 0.02$
No. 10	1: 2.5	$r_t = -0.07 \pm 0.03$
No. 11	1: 5.1	$r_t = -0.08 \pm 0.02$
No. 12	1: 6,540	$r_t = 0.17 \pm 0.02$
No. 13	1: 163	$r_t = -0.13 \pm 0.02$

* Another statement for these results is as follows: If there is really no difference between boys and girls in the choice of each of the several exercises, a difference as great or greater than that found above would arise as a matter of chance in the case of Exercise 5, 1 in 2 times; in Exercise 6, 1 in 7.4; in Exercise 7, 1 in 4; in Exercise 8, 1 in 559,000; in Exercise 9, 1 in 10.4; in Exercise 10, 1 in 3.5; in Exercise 11, 1 in 6.1; in Exercise 12, 1 in 6,541; in Exercise 13, 1 in 164 times.

Because of the position of Exercise 5 among the exercises of the test, because of its brevity of statement, because of the definiteness of the task, and the apparent ease of getting a correct result, there appears to have been no correlation other than pure chance between being a boy rather than a girl and selecting this exercise rather than rejecting it. Between being a boy rather than a girl and selecting Exercise 6, there is a positive, fairly reliable, although low and not particularly significant correlation. For Exercise 7 the boy-to-girl selection-rejection correlation is negative, not significant, and not even reliable. In Exercise 8, the construction problem, we have a relatively high degree of correlation between being a boy rather than a girl and selecting the problem. This correlation is positive and reliable. Moreover, the probability that the observed difference is not a chance difference is expressed by the fact that such a difference in selection as was observed would happen as a matter of chance one in 559,000 times. This difference is probably due to the strength of the manipulation-experimentation interest in the boy. In Exercise 9 we find a low, not particularly significant, though negative and fairly reliable correlation between being a boy rather than a girl and selecting the exercise. In Exercise 10 we find a low, not significant, negative and not particularly reliable coefficient of correlation. In Exercise 11 we find a low, negative, though fairly reliable correlation between being a boy and selecting the problem. In Exercise 12 we find a relatively high degree of correlation between being a boy and selecting the problem. This coefficient is positive and reliable. Here too, there is only a slight possibility that the observed difference can be accounted for as a matter of chance. In Exercise 13 we find a negative, relatively significant and reliable correlation between being a boy rather than a girl and selecting this exercise. The observed difference in selection here can not easily be accounted for as a matter of chance. This may indicate a greater degree of discrimination on the part of boys in avoiding the exercise because of its difficulty.

Abilities of Boys and Girls Related and Expressed in Terms of the Bi-serial r. Up to this point we have used median and mean scores and percentages of perfect scores made by boys and by girls and per cents of boys excelling the median girl as measures by which to compare the sexes in ability to solve originals in geometry. These measures do give standards by which we may make comparisons, yet it is possible to express this relationship in ability to solve originals by a coefficient which can be interpreted as an ordinary coefficient of correlation. Such a coefficient is the bi-serial r, which we will here use to express the relationship between being a boy (rather than a girl) and doing better (or worse) on a problem than at least one half the total group.

This r was first used by Pearson in *Biometrika,* Vol. VII and was later termed the bi-serial r, a designation which it has retained. It expresses, as used here, a relationship between two series of scores through the use of (a) the mean of the distribution having the larger number of measures, (b) the mean of the two series thrown into one distribution, (c) the standard deviation of the combined distribution series, and (d) N, the percentage which the larger series of measures is of all measures combined.

As applied to our present problem of comparing boys and girls in abilities to solve each exercise, we have the formula for the bi-serial r in the following form:

$$r_{b.s.} = \frac{(M_l - M_c)}{\sigma_c} \div \frac{z}{N} \text{ in which}$$

$r_{b.s.}$ = bi-serial coefficient of correlation.
M_l = mean of larger group.
M_c = mean of combined group.
σ_c = sigma of combined group.
N = percentage which the larger group is of the groups combined.
z = a value found from Sheppard's tables corresponding to $\frac{1}{2}(1 + a)$ for N. (See Table II by Sheppard in Pearson's *Tables for Biometricians and Statisticians.*)

The application of this method to the finding of the r for Exercise 5 follows:

EXERCISE 5

In

$$r_{b.s.} = \frac{\dfrac{M_g - M_c}{\sigma_c}}{\dfrac{z}{N}},$$

M_g = Mean of girls (or larger) distribution.
= 9.70.
M_c = Mean of combined distribution of boys and girls.
= 9.88.
σ_c = Sigma of combined distribution.
= 0.3248
N = Percentage larger group is of total.
= 52.7.

Then from Sheppard's tables for: $N = \frac{1}{2}(1 + a) = 0.527$, $z = 0.3980$

$$\text{and } r_{b.s.} = \frac{\dfrac{9.70 - 9.88}{3.248}}{\dfrac{0.3980}{0.527}} = \frac{\dfrac{-0.18}{3.248}}{0.7552} = \frac{-0.055}{0.7552} = -0.073.$$

$$r_{b.s.} = -0.07.$$

The correlation between the successes shown by boys and girls in solving this exercise is:

$$r_{b.s.} = 0.07 \pm 0.03.$$

For the computation of P.E.$r_{b.s.}$ see page 111.

The bi-serial r's for the scores made in the several exercises by boys and girls, computed as in the case of Exercise 5 above, are given in Table XV.

Summary of Data on Correlation in Abilities. The r's for the correlation in abilities may appear small, yet from our general knowledge of sex differences as to ability in intellectual pursuits we have little ground for expecting the study to show large sex differences in abilities involved in the solution of these exercises. The differences as they appear are shown by the coefficients of correlation.

For convenience in reference, the bi-serial r's obtained as in the case of Exercise 5 above, are here set down in connection with the tetrachoric r's previously computed.

TABLE XV

Exercise	No. 5	No. 6	No. 7	No. 8	No. 9	No. 10	No. 11	No. 12	No. 13
Bi-serial r	0.07	0.12	0.14	0.38	0.33	0.22	0.18	0.15	0.32
Tetrachoric r	0.00	0.08	−0.09	0.32	−0.09	−0.07	−0.08	0.17	−0.13

An examination of the bi-serial r's above with respect to the size of their probable errors (see column K, page 114) shows a high degree of reliability for each coefficient except the first, for reliance may be placed on a given correlation coefficient if the coefficient equals four times its probable error.

It should be noted also that all of the coefficients are positive thereby presenting a really dependable piece of evidence that being a boy (rather than a girl) leads to doing at least a little better work in solving originals. Let it be noted also that the highest positive *r* (tetrachoric *r*) between being a boy and selecting the exercise (see Exercise 8) goes with the highest *r* (bi-serial *r*) between being a boy (rather than a girl) and excelling the median girl in the solution of the exercise. The nearest zero *r* (see Exercise 5) between being a boy and selecting the problem goes with the nearest zero *r* between being a boy and succeeding. Again, the greatest negative *r* (see Exercise 13) between being a boy and selecting (this means the highest positive *r* between being a boy and rejecting) goes with one of the highest *r*'s between being a boy and succeeding in the solution of the problem selected.

The evidence on Exercises 9, 10 and 11 is not so conclusive, however, for in these exercises low and almost chance correlations in selection between being a boy rather than a girl go with significant though low correlations in ability to solve the exercises.

The fact that five of the nine bi-serial *r*'s are below + 0.20 and that but three *r*'s are above + 0.30 leads to the conclusion that while sex differences in abilities do exist, yet these differences are so slight as to give scant justification for segregating boys and girls into class groups, or for reorganizing the subject matter of plane geometry so as to present distinct types of work to each sex group. This conclusion is borne out by the following facts:

In two of the three *r*'s which are above + 0.30, the percentages of pupils selecting the exercises are 17.9 and 18.5 respectively. Since we are able in these exercises to contrast the abilities for but one pupil out of five or six there is little ground for assuming that there is considerable sex difference in ability to solve originals in geometry. There remains, then, but one exercise out of the nine considered in which the study shows a really significant sex difference, that is, Exercise 8. Here the strength of the manipulation-experimentation interest in boys appears in a decided preference for the construction exercise, and also in a superior ability to solve the exercise. Having determined the type of exercise (the construction exercise) in which the sex difference appears greatest, we shall now utilize our data on the analysis of that exercise (see pages 12–15) to locate within the processes involved in the solution of that exercise just those processes which account for the observed sex difference.

Location of Difficulties for Girls in Solving Construction Exercises. In the expressed preference for one exercise rather than another, the study has shown that the highest correlation between being a boy (rather than a girl) was found in the data on the construction exercise. Likewise on this same exercise the highest correlation in ability to solve an exercise was found.

The data on the analysis of this problem show that of the 503 boys and 559 girls who attempt the exercise, 455 boys and 466 girls note that by the conditions of the exercise, a circle and a line outside the circle are given or fixed. From these data, by means of Pearson's square contingency coefficient (a coefficient interpreted here as other coefficients of correlation used in the study except that a perfect correspondence for a four-fold grouping is indicated by 0.70), we are able to find the correlation between being a boy, rather than a girl, and noting that the circle and the line outside of it are given or fixed. The method of obtaining this coefficient is in part identical (obtaining of X^2) with that used in obtaining the probability of association. (See pages 53–56.) It is here given:

Interpretation of Given Elements	Boys	Girls	Total
Correct	455	466	901
Incorrect	48	132	180
Total	503	578	1081

Here $X^2 = 34.78$ which leads to the

$$\text{Coefficient of Contingency} = \sqrt{\frac{34.78}{1115.8}} = 0.176.$$

From the above it follows that there is a distinct but not high association, 0.18, between being a boy and noting correctly the given elements of a construction exercise. Apparently, then, most girls need to exercise more care than boys in identifying the given elements in an exercise of this type.

If we consider only those pupils who correctly interpret the given elements of the exercise (455 boys and 446 girls), we find that 66 boys and 161 girls do not note that in the solution it is worth while to drop a perpendicular from the center of the given circle to the given line, or who, seeing that it is worth while, fail in their attempts to carry out the construction of the perpendicular. We then compute the contingency as follows:

	Boys	Girls	Total
Noted	389	285	674
Not noted	66	161	227
Total	455	446	901

Here $X^2 = 55.7$ which leads to the

$$\text{Coefficient of Contingency} = \sqrt{\frac{55.7}{956.7}} = 0.241.$$

From the above it appears that for those boys and girls who correctly interpret the given elements in the exercise, the contingency between being a boy (rather than a girl) and constructing or attempting to construct a perpendicular from the center of the given circle to the given line is 0.241. The contingency coefficient of 0.24 indicates that the association between being a boy and seeing the need for the perpendicular does exist. The relationship, however, is not high.

In the use of the "mid-perpendicular" method of constructing a line perpendicular to the line drawn at the point of the intersection with the circle, the data show a contingency of $+0.243$ between being a boy and performing this construction. Here again we see that an association does exist, though we do not have a high degree of relationship.

In actual compliance with the requirement that all construction lines be clearly shown, the contingency between being a boy and meeting the requirement is 0.143.

In the evident use of the theorem, "Two perpendiculars to the same straight line are parallel," as an idea fundamental to the correct analysis of the problem, the contingency between being a boy and calling up the idea is 0.13. In this item as in the item immediately preceding, the association is low. It is in fact so low in each of these cases that one could not safely base a judgment as to a sex difference in a method of procedure upon the difference found.

In the use of the theorem, "Two points each equally distant from the extremities of a straight line segment determine a mid-perpendicular to that segment,'' as an idea necessary to the understanding of the means of constructing a perpendicular to the given line from the center of the given circle, there is a contingency coefficient of 0.03 between being a boy and calling up this idea. A coefficient of this size indicates that no relationship exists other than that which can be accounted for on the basis of pure chance.

The results obtained above show that a real sex difference does exist in the determining of what is and what is not given in the statement of the construction problem. Moreover, such a difference does exist in abilities to see that in determining the point of tangency of the required line a perpendicular must be dropped from the center of the given circle to the given line. The size of each coefficient indicates that in each case the difference will not always be found. This difference, though probable, is by no means certain. Between being a boy rather than a girl and using the ''two point'' construction of a perpendicular to a line segment, the coefficient 0.03 shows clearly that in this matter there is almost certainly no sex difference.

From the above data it seems clear that in teaching the methods involved in the solution of construction exercises, more care must be taken with girls than with boys. The need for this greater care is particularly manifest in distinguishing between the given and the required elements of the problem. The first steps of the construction problem evidently present difficulties to girls more often than to boys. Good teaching of this type of work should recognize the difference observed in this study.

The method here used is an attempt to run down mental sex differences found in the solution of this construction problem to the elementary intellectual processes involved. The analysis has definitely located the difference in certain of the following mental operations and not at all or slightly in other operations:

1. In noting that the circle and the line outside of it are given or fixed by the conditions of the problem.
2. In nothing that it is worth while to drop a perpendicular from the center of the circle to the given line in order to locate the point of tangency of the required line.
3. In the use of the ''mid-perpendicular'' method of constructing a line perpendicular to another line.

4. In the actual compliance with the requirement that all construction lines be clearly shown.
5. In the evident knowledge of the theorem, "Two perpendiculars to the same straight line are parallel."
6. In the use of the theorem, "Two points each equally distant from the extremities of a line segment determine a mid-perpendicular to that segment."

There is certainly no sex difference in the operations involved in (6) above. Similarly the mental operations listed as 1–6 inclusive, pages 12 and 13, are neutral so far as sex differences are concerned.

An adequate explanation of why such differences take the form here shown must be left to the introspectionist. It must suffice here to establish a method and supply a small amount of data showing mental sex differences in fairly elementary processes.

Conclusions. The general conclusions of this chapter are:

1. The most pronounced preference for one type of exercise rather than another is that shown by boys in selecting the construction exercise. Here the correlation coefficient between being a boy and selecting the exercise was + 0.37. This is probably due to the strength of the manipulation-experimentation interest in boys. Girls also select this exercise more often with one exception than any exercise involving a numerical solution or a proof.

2. The several measures of ability used all show that the male sex slightly excels the female in solving original exercises. There is a positive though low correlation between being a boy and excelling a girl in solving each original considered. The study shows, however, that the differences between the sexes in the abilities here observed (61 per cent of the boys equal or excel the median girl) are far outweighed in importance by the variability within each sex group. On the score of abilities, therefore, no evidence is here found on which to base a claim for segregation of the sexes for the study of plane geometry.

3. The study shows, moreover, that the amount of overlapping of the sex groups in abilities to solve numerical exercises is approximately equal to overlapping shown in abilities to prove exercises which involve demonstration. There seems, therefore,

to be no good reason for assigning numerical exercises to one sex group and exercises involving demonstration to the other.

4. The study shows that among pupils who having elected to be examined upon certain exercises pass the examination as a whole, the variability of boys is somewhat less than of the girls.

5. The range of boy-to-girl variability in the solution of numerical exercises is approximately equal to the range in variability in solving demonstration exercises. This indicates that one type of exercise is not much better suited to either sex than the other type.

6. The study shows that in the solution of the construction exercise girls have greater difficulty in distinguishing between given and required elements than do boys. The method here used in presenting this fact, the contingency coefficient, indicates a method by which mental sex differences may be located in fairly elementary processes.

CHAPTER IV

STEPS IN THE THOUGHT PROCESSES INVOLVED IN SOLVING ORIGINALS

Purpose of the Chapter. It is the purpose of this chapter to show first that the solution of originals in geometry involves thought processes of the "trial and success" type, and, second, that success in solving original exercises is made more certain through an orderly procedure from the known to the unknown.

This procedure seems from the study to be expressed under the five following heads:

1. Identification of the given and required elements (facts or relations) of the problem.

2. Search for direct outcomes of these elements and for groups of elements which involve given and required elements in known relations.

3. Selection of certain groupings of given and required elements as worthy of closer investigation in the attempt to pass from the given facts to the required outcome.

4. Utilization of groupings of elements selected as in (3) above through: *a*) Testing of a suggested method of reaching the desired outcome. *b*) Testing as in *a*) above and rejecting the method suggested, or *c*) Creating new groupings of elements through drawing construction lines, etc., and then testing such combinations as in *a*) and *b*) above.

5. Verification of the results of a completed solution through: *a*) Syllogistic organization in exercises involving proof. *b*) Interpreting and checking results in problems involving numerical data. *c*) Discussion of limiting conditions and proof that the construction is correct where solutions involve construction.

Thought Process Employed in Solving an Original Exercise. Granting that the individual is in possession of knowledge necessary to the solution of the exercise, it appears from the study that the particular reactions he will make depend on the force with which certain elements of the exercise strike him. That the reactions made and the solutions attempted are not always the most economical is evident from the study. Indeed no one familiar with the working mind would expect the pupil to think in terms of the syllogism. The following quotation from Woodworth's *Dynamic Psychology* bears on this very point:

> But it is now recognized that the formal syllogism is by no means a psychological picture. It is a check which can be applied to a completed act of reasoning, to detect possible errors. . . .
>
> This can easily be demonstrated by the solution of what are called "originals" in the teaching of geometry. It is true that the regular propositions in the geometry books are set down in syllogistic form, with an orderly procedure from the known to the unknown. But it is safe to say that these propositions did not originate in this well-ordered form and this can be demonstrated, or at least be made highly probable, by observing how an original is solved. One can not go straight forward in an orderly manner —if one could, not originality but habit would be in play. One starts with a problem, and explores about like a rat in a maze or a cat in a cage, trying this and that as one notices one after another feature of the problem, till finally a good clue is got, the essential elements of the problem are discovered and the appropriate premises are recalled—after which trial and error process, one can remodel the reasoning into syllogistic form and thus check its correctness. Reason thus proceeds from the unknown to the known. . . . It would be easy, no doubt, to start with the known, but the question would then be, whither to proceed. We need a goal. The goal is the unknown, which comes first in reasoning, as a goal, to be sure. Reasoning is first of all a tendency toward the unknown, and next a finding something known from which to proceed. The unknown, strange, baffling situation must somehow be made to yield something that is known, by means of which the unknown can be mastered.

The list of the several steps used by pupils in solving the originals here examined (see pages 6–29) shows that to different individuals the same printed exercise presents very different combinations. The potency of the elements in the problem is determined by the care with which the problem is examined and by the individual's mental equipment.

No amount of care exercised later in the solution can make up for a careless examination of the problem, or for the neglect of one essential element. Even though this process of examination is similar, as Woodworth says, "to the action of a cat in a cage or a rat in a maze," yet this sort of reaction should follow and not precede a most painstaking examination of each element of the situation with its immediate outcome. The process of exploration is indeed essential, but should never be begun until the given elements of the problem are clearly comprehended and the goal clearly defined.

That individual is certain to be most original or most responsive to a situation who is most familiar with the kind of material or facts used. Hence the pupil who is familiar with his definitions and previously proved theorems is prepared to deal successfully with those new situations in which these facts are essential elements. Even a genius does not fail to consider the essential elements of his problem. To such a one, each element presents a variety of possibilities. It is because of this wide range of possibilities suggested that successful selection of a worthwhile outcome is more probable in his case. This statement is borne out by a quotation from Thorndike (*Teachers College Record,* November, 1916): "In fact, a good definition of intellectual independence is 'reasoned dependence'."

If one could get a complete record of the thought processes of each of many pupils as they work at originals in geometry, one would no doubt have data which would show that thinking in this field is of the "trial and success type." The attempts at solution both successful and unsuccessful as recorded in this study (see pages 6–29) show certain tendencies and methods to be more economical and on the whole more successful than others.

The first step after a careful reading of the statement of the problem is and must of necessity be a clear identification of the elements of the situation. In this particular, the data collected from the study show that most pupils seemingly identify with some care the several elements in most problem situations, both as to what is given and what is to be found or demonstrated. It is in the next step that the records made in this study show particularly uneconomical methods.

While we can secure from the papers no mental picture showing the thought processes of each individual pupil, yet the varied attempts of the many indicate in a measure what does take place in the thought processes. It is evident that even though all read the same problem and note the given elements and the desired outcome, yet the methods of using the given elements so as to attain this outcome show "varied reaction and multiple response."

To some pupils one element in the problem seems potent and to another pupil another element. If the reasoning processes involved require more than one step, then the end sought is not the single controlling factor in taking the first step in the solution. In such cases the greater familiarity on the part of the pupil with theorems which involve a fact or relation suggested by one element in the solution determines the entire method of procedure, at least for a time.

With those who fail in the solution or with those who succeed through the use of an uneconomical method, the end sought seems to have greater potency than the known elements. Probably a common thought of all such is, "Perhaps I can prove in this (a suggested) way." In an involved situation such as is found in Exercise 12, the method adopted for proof too often involves the drawing of construction lines, which, when drawn, because of their recency, suggest new combinations of elements, thus practically creating new situations. Just here we find "the cat in the cage type of thinking."

If this type of thinking leads to a successful outcome, the essential elements in the given situation must in any case be identified and used. The records made show that pupils do identify more or less carefully the given elements, yet with those who employ long methods, the elements in the desired outcome seem over potent in determining the procedure—a procedure which too often involves the drawing of unnecessary construction lines. The desired outcome should be a potent element in the solution, but its potency should not be so great as to determine the method of procedure more or less independent of the given data. In that case, we have mere random experimentation.

When, however, this desired outcome is held in mind as a point to be reached or a goal to be attained after using all the given elements in the situation, a use made at or near the very outset, then economical thinking seems more certain.

The original exercises of this study may be classified as those involving (1) a numerical solution, (2) a construction, and (3) a proof. The thought processes employed by pupils in their final attempts at solution of one exercise of each type are presented in the diagrams which accompany the discussions of the methods used in solving exercises of the above types.

Thought Processes Employed in Solving Exercise 12, Involving Demonstration. The several methods used by pupils in solving Exercise 12 are expressed briefly and are shown in Diagram 11.

Step I. The solution as made by the pupil must of necessity begin with a more or less careful reading of the exercise with an accompanying identification of the given and required elements. Since this step must of necessity come first, we have designated it here and shall refer to it as Step I. This step must be taken by all who attempt to solve the exercise. In cases where the facts of geometry are to be applied, we may have first a realization on the part of the pupil that a difficulty or exercise is at hand and that this exercise requires a solution. In the solutions as worked out by the pupils during their study of plane geometry, however, the exercise is either dictated by the teacher or read from a text book. In either of these cases the first reaction on the part of the pupil will be an identification of each of the several elements (lines, angles, etc.). This step usually includes the sketching of a figure, the parts (or relations of parts) of which represent the given and required elements.

That the pupils take this step first is not verified by the data at hand, for it should be recalled that only the final record of work done by pupils has been examined. It seems probable, however, that the pupil will begin his solution by reading his exercise word for word rather than by searching for what is given and for what is to be proved. In either case there is an identification of the elements of the exercise.

Step II. As in Step I above, the evidence offered by the papers examined is not sufficient to reveal the initial method of attack used by the pupil. It is obvious, however, that he must either carefully examine the given elements to determine their direct outcomes, as in Step II of the diagram, or he must hit upon a certain grouping of elements in which the desired outcome is more or less directly involved. That situation will be most potent which involves most known elements or elements potent because of recency of use, etc. The pupil is often impressed by a particular combination of elements, each element (or group) of which he examines further to get a better understanding of the relations of the given elements. A new situation

DIAGRAM 11. STEPS IN SOLUTION OF EXERCISE 12

STEP I — *Identification*

Identification of given and required elements. (Total 1101, * N, 6.)

STEP II — *Search*

A. $MN \parallel BC$ → (Total 863)
⊥s to l, 577
alt. int. ∠s, 74
cor. ∠s, 207
sup. ∠s, 15

B. MN not recog. $\parallel BC$. Sim. △s are or can be formed. → (Total 183)

C. No record of outcomes of any kind. (Total 6) → Did practically nothing with problem. Total 6.

STEP III — *[Selection]*

STEP IV — *Utilization*

A_1 (Method) $AM = MB$; then $AN = NC$.
*(Report) (C, 339; U, 4; N, 74; W, 4.)

A_2 Line ∥ base △; divided sides prop.: $AB = BC$; $AN = NC$.
(C, 267; U, 5; N, 4.) BC

A_3 Drew $AD \parallel MN$; $AD \parallel MN \parallel NC$; 3 ∥s cut off = seg.; $AN = NC$.
(C, 114; U, 1; N, 75; C, 79; U, 1; W, 109.)

A_4 △s AMN and ABC sim.; cor. sides prop.; $AN = NC$.
(C, 48; U, 18; W, 3.)

A_5 $\angle 1 = \angle 2$, $\angle 1 = \angle 4$, $\angle 2 = \angle 3$, hence $\angle 3 = \angle 1$, $AN = NB = NC$.
(C, 6; W, 2.)

B_1 Drew $CX \parallel AB$; prod. MN to X, △ AMN = △ NXC; $AN = NC$.
(C, 23; U, 2; N, 1; W, 1.)

B_2 Drew $NH \perp BC$; proved △ AMN = △ NHC, $AN = NC$.
(C, 6; W, 2.)

B_3 Rt. △s AMN and ABC sim. to get sides prop.; $MN \parallel BC$, then (A_1).
(C, 1.)

B_4 Formed rt. △s AMN and MNB; BNH and HNC; then $AN = NC$.
(C, 36; U, 31; W, 2.)

B_5 $NHB = NHC$; $NB = NC$; ANB isos.; $NB = NA$; then $NC = NA$.
(C, 32; U, 24; W, 2.)

B_6 Put ⊙ about △; proved $AN = BN = NC$.
(C, 2; U, 1; W, 3.)

B_7 Formed rectangle $ABCD$; diag. bisect. $AN = NC$.
(W, 9.)

B_8 Formed rt. △s NHC and NMA, proved =, then $AN = NC$.
(C, 17; U, 19; W, 4.)

B_9 Assumed $AN = NC$.
(W, 8.)

Did practically nothing with problem. Total 6.

* Key:
C = Correct.
U = Unsatisfactory.
N = No use made.
W = Wrong use.

STEP V — *Verification*

Syllogistic organization not required here.

grows out of a search for the direct outcomes of the given elements with a recalling of previous combinations (theorems used).

From the data at hand, it seems that the selected situation does grow out of a careful examination of the several combinations of the elements. The situation thus selected may and should involve all of the given elements or lead to a situation which will involve all of them. The chance situation hit upon and utilized often does not involve all given and required elements. In the solution by pupils the study shows that 863 or 82.5 per cent were impressed by the fact that lines *MN* and *AB* were parallel while only 183, 17.5 per cent, were not so impressed.

The combinations of elements which lead to the conclusion that lines *MN* and *BC* are parallel or those combinations which were called upon in support of the inference are indicated in the diagram. There seems to be evidence that the idea "*MN* and *BC* are parallel" did not occur to the 183 (17.5 per cent) pupils who follow lines of thought indicated in Step II-B. The solutions of these pupils are more indirect and involved. In the course of such solutions recourse must sooner or later be had to some of the direct outcomes of the given elements. It seems reasonable, therefore, to conclude that the solutions will be more simple if the pupil searches for (1) the direct outcome of each element given and (2) in a general way the implications of the desired outcome before picking out or accepting a certain situation (group of elements) as the best means of reaching the desired goal.

Step III. Either before carefully examining the several elements to see their direct outcomes or after such an examination, the pupil *selects* a grouping of elements which contains the given elements and the elements of the outcome or one which gives promise of leading toward a grouping which does contain such elements. Along with the selection there goes a decision to test the relations involved to see if the desired outcome may be attained.

This selection and decision to test a grouping of elements does seem to follow rather than precede (77.7 per cent of cases) the examination of the given elements. Where this selection does seem to precede, it may well be based upon certain apparent results (right angles are formed, etc.) rather than upon direct outcomes of elements grouped.

In any case, there is at some point (which we term here Step III) a decision to test a certain situation (combination of elements) in order to attain the desired outcome.

Step IV. Pupils who react successfully to a problem do test or utilize a selected or chance grouping to attain the goal desired. The several groupings depend both upon the situation as a whole which is selected and upon the potency of the several elements therein. Such utilization, therefore, consists in *a*), the testing of the combination first selected as probably useful, which in cases where all known and required elements are grouped in known relations results in a successful solution; *b*), the testing and rejection of the situations selected and tested as in *a*) but where the solution is not reached; or *c*), the creation of new situations (by new groupings, use of construction lines, etc.) which are tested as in *a*) or tested and rejected as in *b*).

This step, which is here termed "utilization," depends wholly upon the nature of the situation hit upon as probably useful in Step III. The varied nature of this process for a given exercise is seen by an examination of the foregoing diagram.

Since the examination papers show only the last situation tested (in many cases imperfectly tested), reports can not be made on solutions tested and rejected. The data do show that those pupils who are successful in their solutions select such situations as involve all of the given and required elements of the problem, and that those situations which in their initial stages involve all known and required elements result in less complicated solutions.

Step V. The final stage of each successful solution of a problem involving proof is the *syllogistic organization* of the statements used and supported by evidence in the course of reasoning by which the pupil passes from the statement of the problem to the ultimate conclusion. This more or less perfect organization is in content, of course, determined by the situation (combination of elements) selected, utilized, and found useful. The papers examined offered only this organization. It is from this organization that we have been compelled to judge the situations selected and utilized.

Thought Processes Employed in Solving Exercise 7, Involving a Numerical Solution. Although the pupils were not required in their solutions of Exercise 7 to set down the theorems used in the several steps taken, yet from the equations used and numerical results obtained from the several steps which had evidently been taken, it is possible to locate the theorems used. In this manner the thought processes of the pupils are traced through their solutions. Since verification of results is not specifically required, this process, if performed at all, is carried out mentally or on scratch paper and is not recorded on the papers examined.

The several lines of thought used by pupils in their solution of Exercise 7 are shown in the accompanying diagram.

Step I. From the diagram (on page 75) it is evident that practically all of the pupils examined note that they have given: a triangle of sides 10, 16, and 20 feet and a line parallel to the base cutting off 2 feet from the lower end of the 10 foot side.

If from finding or attempting to find two segments of the side 16 feet, we may measure the actual identification of the fact that two segments of the side 16 feet are required, the data show that 85.0 per cent do clearly identify this essential element. On the contrary, 12.4 per cent give no evidence of having noted the fact that the length of the line parallel to the base is required.

The actual noting of what is given and what is required is here called Step I. This step is essential for all who solve the problem.

Step II. The data show that *only* 97 per cent of the pupils note that the two parts into which the parallel divides the 10 foot side are 2 feet and 8 feet.

All who attempt the actual solution of the exercise see that to reach one or more of the desired outcomes use must be made of a proportion which is found by relating numerical measures of the corresponding sides of similar triangles or of theorems involving proportional segments. The search for a combination of elements which relates given and required elements in known relations is then begun.

Step III. The search for situations as in Step II above leads sooner or later to the selection of a certain combination of elements as sufficiently worthwhile to justify its use as a means

DIAGRAM 12. STEPS IN SOLUTION OF EXERCISE 7

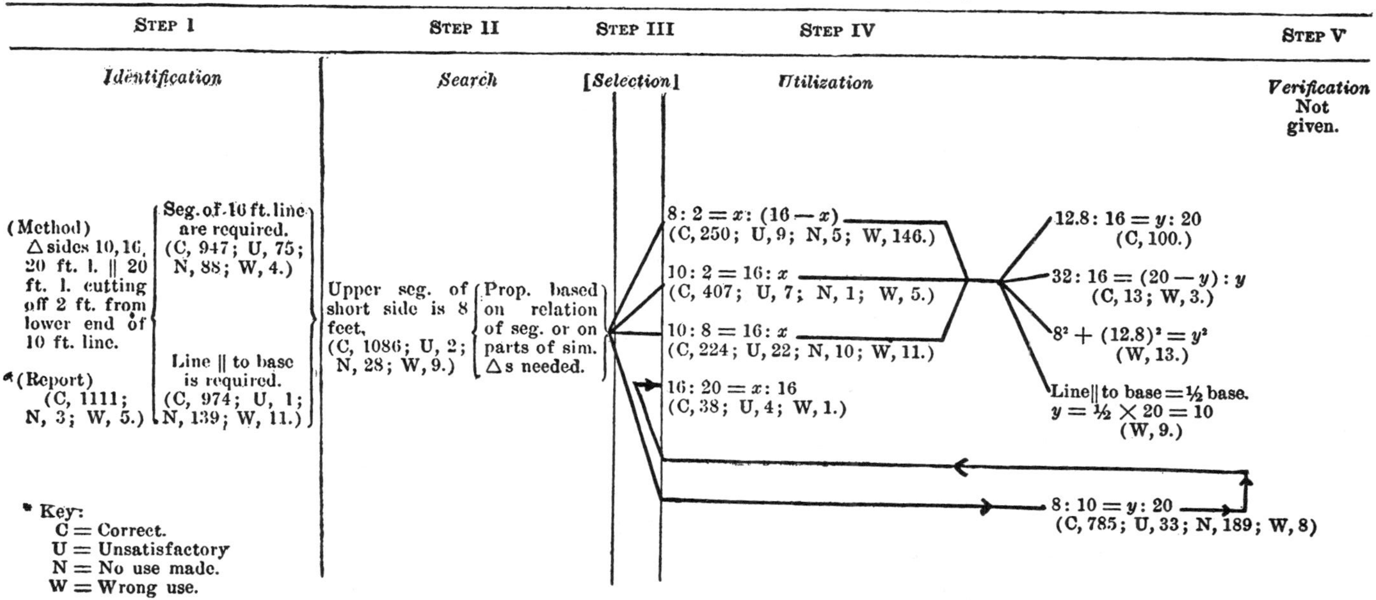

of securing one of the desired outcomes. The particular situation hit upon as useful is determined by the potency of the several elements in the exercise as stated and by the past experience of the pupil in dealing with situations more or less closely related to this exercise. If the pupil has had considerable experience in dealing with proportions which come directly from relations of corresponding parts of similar triangles, a situation of this sort will be here recognized. If the pupil has recently or frequently and successfully used a theorem which relates segments of lines, the possibility of doing likewise here may suggest itself.

From among the pupils who decide to find first the segments of the side 16 feet rather than the length of the line parallel to the base, the data show that pupils hit upon the possibility of using a relation between segments of lines more often than upon one involving relations of sides of similar triangles. This is probably due to the use of the word "segments" in the exercise as stated. The ratio of those using the theorems of proportion directly related to similar triangles to those using theorems directly relating segments of lines is 3 : 1 approximately. The word "segments" then seems not to be a potent factor in leading the pupil to use the theorems involving relations of segments.

Another interesting fact appears in the same connection. As indicated in the diagram there are two ways in which relations of segments may be utilized, one involving the proportion $8:2 = x:(16-x)$ and the other, $10:2 = 16:x$. The first relation is used by 410 pupils as against 421 using the second. The first is based upon the use of a general theorem, while the second is based upon a relatively unimportant corollary. The method involving the use of the relation expressed by the corollary is selected, however, because three of the four terms of the proportion are given in the figure as drawn, while but one term of the first proportion is specifically mentioned in the exercise as given.

Those pupils who search out the direct outcomes (as in Step II) of their given elements will have the parts $(8, 2, x, 16 - x)$ which when expressed and considered in connection with parallel lines give promise of suggesting the general theorem which directly relates the given and required segments only. The discussion given here indicates at least some of the factors which are potent in leading the pupils to select a certain situation as probably useful in reaching the desired goal.

Step IV. After having cast about as in Step III above for a situation involving known and unknown elements, and having selected a certain combination of elements as there described, the pupil proceeds to test his selected combination of known and unknown elements in order to secure the desired outcome. If the first grouping selected yields this outcome, the original work of solving this part of the exercise is complete. If the desired outcome is not so obtained, then he re-examines his given and required elements to see if all are used and if they are properly related to each other. If this examination suggests the use of a new combination or relation, it is tried out. It is accepted if it yields the desired outcome and rejected if it does not. If time permits, the pupil then re-examines his exercise to seek a new relation of all the given and required elements, possibly drawing construction lines, thereby creating new situations which give promise of suggesting new and probably useful relations. Each situation suggested is tested in turn as above until the desired conclusion is reached or the solution put aside.

As is seen in Diagram 12, the results of the solution of the first part of the exercise, although not needed, are often potent in determining the method to be used in solving the second part (the length of the line parallel to the base). This is true here in 13.4 per cent of the cases in which pupils obtained the segments of the 16-foot line.

Again, 43 pupils out of the total 1119 or 4 per cent recognize first a situation which makes possible the solution of the second part of the exercise (to find the length of the line parallel to the base) and then instead of referring to the original data these pupils use the new elements in solving to find the segments of the 16-foot side,—the first requirement of the exercise as stated.

Since the papers examined showed only the final work of the pupils, data are not at hand to search out the false trials or the testing and rejection of the situations. From the final data we are able to see, however, just what facts and relations pupils in groups possessed or recalled and judged most useful.

Step V. Since the exercise as stated did not require a checking of the answers obtained, the papers did not show that the process was recognized as essential. If performed at all, it was performed mentally or on scratch paper and was not recorded in the examination papers as submitted.

Thought Processes Employed in Solving Exercise 8, Involving Construction. Through references to Diagram 13, which is based upon records made of the pupils' work in solving the construction Exercise 8, it is evident that practically all pupils do clearly identify the fact that a line is to be constructed which is to be parallel to a given line. It is evident, however, that 16.6 per cent of the pupils fail to note that the line and the circle are fixed or given before the construction is begun. There is here a careless identification of the given elements which in 11.9 per cent of the cases results in a construction based upon a wrong assumption. Again 0.6 per cent of the pupils disregard the fact that the given line does not intersect the given circle. This, too, results in a construction based upon a wrong assumption. Again, 17.3 per cent disregard the fact that construction lines must be clearly shown,—a specific direction.

Step I. It is evident that the solution of a construction exercise must begin with the careful identification of all given and required elements. The percentages quoted above show that pupils need more drill on this feature than is provided for in geometry as taught. The record, made on the use of construction lines, shows that 82.7 per cent of the pupils note carefully that all construction lines should be clearly shown. Evidently pupils do about as well here in discriminating between what is and what is not given as they do in following a clear-cut instruction which requires that all construction lines be clearly shown.

Step II. After noting more or less carefully the given and the required elements, the successful pupils begin to note the direct outcomes of the given elements. This process is really a continuation of the identification of the given elements. Here the pupils note that a tangent which is desired is perpendicular to the radius drawn to the point of contact on the circumference, lines to be parallel may all be drawn perpendicular to the same straight line, etc. Such facts are or should be noted in the course of the solution by all of the pupils.

DIAGRAM 13. STEPS IN SOLUTION OF EXERCISE 8

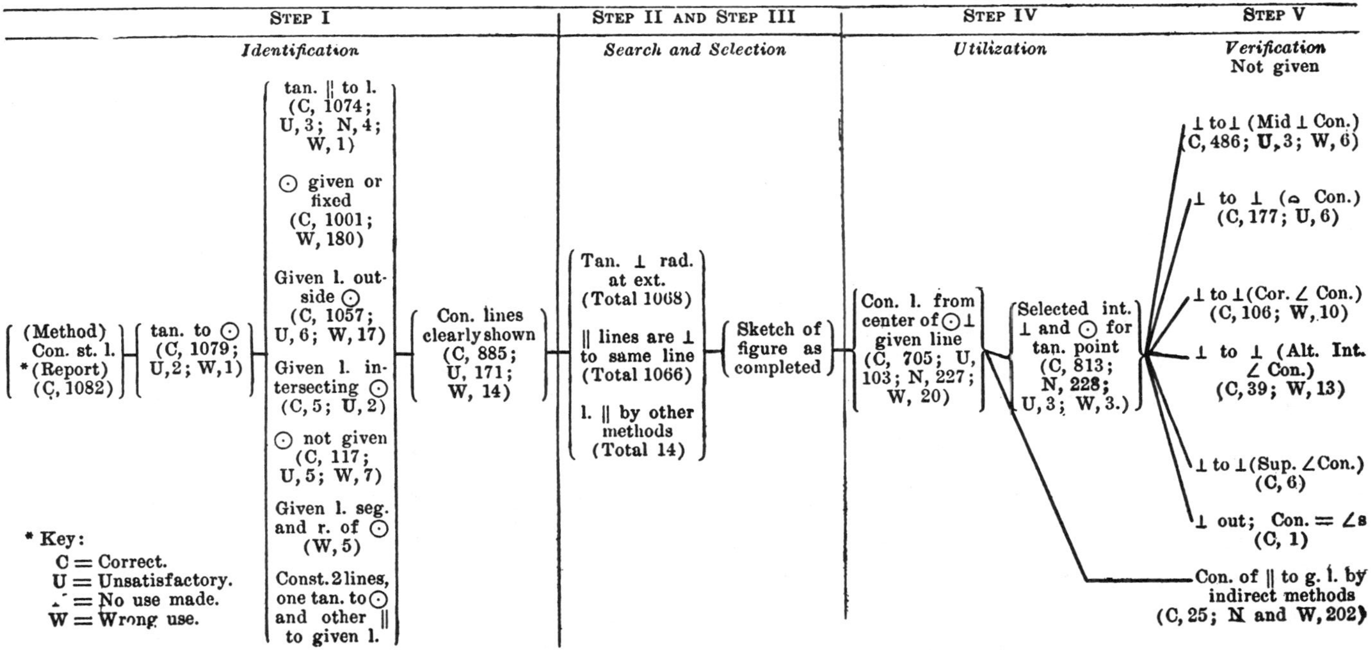

The study does not show, yet there seems to be some ground for assuming, that the successful thinker makes at the outset a mental picture (or better a rough sketch) of the given and required lines as they will appear when drawn.

Step III. A search is then begun for such means as will make possible the construction desired. In 96.6 per cent of the cases the pupils decide that a line must be constructed from the center of the circle perpendicular to the given line and 75.6 per cent decide that at the point or points of intersection of the line last drawn and the circumference of the circle a perpendicular to that line must be drawn. Others (24.4 per cent) decide to draw a parallel to the given line which will also be tangent to the circle, by striking arcs of circles whose centers are on the given line, the radii of such circles being in each case equal to the distance of the circle from the line, or by using one of several other indirect methods.

Step IV. The actual construction of the line from the center of the circle perpendicular to the given line is attempted by 75.6 per cent of the pupils, and at the point of intersection of this line and the circle the construction of a perpendicular is begun.

The potency of the several methods of constructing this perpendicular is indicated by the frequency with which these methods are used in the course of the solution. These percentages are: mid-perpendicular construction, 62.7 per cent; semi-circular construction, 17 per cent; corresponding angles construction, 13.4 per cent; alternate interior angle construction, 6.1 per cent; etc. More than a score of different methods of construction were observed.

Step V. Since the work of the pupils on the papers examined concludes with the actual construction of the required line, we have no evidence on the methods used by pupils in proving their constructions correct or on their consideration of the limiting conditions of the exercise.

The data at hand show that but few pupils examined the exercise critically enough to see that two lines satisfy the conditions of the exercise. The process, however, may have been attempted by others.

The examination of the work of pupils on this construction

exercise reveals certain steps as necessary in the course of the complete solution. These steps are:

1. The identification of the given and required elements, including a rough sketch or mental picture of the construction as completed.
2. The search for the direct outcome or given elements and for groupings which will be useful in the solution.
3. A selection of a method of carrying out the construction.
4. A testing of the method selected as probably useful in the construction.
5. A verification of the construction through a proof that the construction is correct and through a discussion of the limiting conditions of the problem.

Conclusions. Formulation of Steps in the Thought Processes. The work done by pupils in solving originals of the demonstration, numerical solution, or construction types suggests a formulation of the steps of the essential solution processes involved under the following five heads, which formulation is offered to teachers of geometry as an outline for frequent reference on the part of both teacher and pupil:

I. Identification of
   A. All given elements (facts or relations).
   B. Elements of the desired outcome (facts or relations to be proved).

II. Search for
   A. Direct outcome of each given element considered alone, and apparent implications of the desired outcome.
   B. Situations (combinations of elements) which do or would involve known elements, or known elements and the desired outcome.

III. Selection of a certain situation because it involves most known elements or the potent elements which give promise of being useful in reaching the desired goal.

IV. Utilization of a situation and its outcomes as a new situation, etc., leading toward the desired outcome. This step includes or may include:
   A. The testing of a selected situation and its outcomes to reach the desired outcome, or

B. The testing and rejection of a selected situation and its outcomes as of no worth in reaching the desired outcome, or

C. The testing and rejection as in B above, followed by drawing of construction lines, providing thereby a new situation which is then tested as in A or B above, etc.

V. Verification of conclusions in:

A. Problems involving numerical data through interpretation of and checking of results.

B. Problems involving construction through consideration of limiting conditions and proof that the construction is correct.

C. Problems involving proof through syllogistic organization.

CHAPTER V

SUGGESTIONS FOR IMPROVED METHODS OF TEACHING GEOMETRY

Purpose of the Chapter. It is the purpose of this chapter to point out to teachers of geometry the necessity for:

1. Recognition of individual differences of pupils.
2. An exact statement of each exercise presented to pupils.
3. Teaching pupils to identify the given and required elements of the exercise.
4. Teaching pupils to proceed in solution by economical methods.
5. Teaching pupils to couple initial outcomes with unused given elements.
6. Class room drill on essential theorems.
7. Teaching pupils to interpret and to examine results to see that all requirements of the problem are satisfied by the results as obtained from the solution.

On the above points the study shows that:

1. Wide differences in abilities do exist even among those who pass the examination as a whole. The differences among pupils should be recognized in teaching that success for each pupil may be made probable.

2. More care than is commonly exercised should be placed on the exact statement of the several exercises of an examination.

3. In the lesson assignment on new and really difficult originals the teacher should direct the class in identifying the several given and required elements of the exercise.

4. Economical methods are not commonly utilized by pupils.

5. In lesson assignments pupils should be made conscious of the need of coupling initial outcomes with unused given elements.

6. Continued class-room drill upon fundamental theorems is necessary. Even after a year of study the right triangle relation is not sufficiently well learned by all pupils to guarantee its correct use where needed.

7. Pupils should be taught to designate the character of and to interpret and to check the results of each solution. That such

habits of dealing with solutions are not commonly formed is obvious from the study.

Recognition of Individual Differences. One purpose of teaching is that of so manipulating the elements of a situation in which the pupil is confronted by subject matter that certain outcomes are made more probable than other outcomes. To do more, that is, to make certain outcomes inevitable, is to rob the pupil of the satisfaction of mastering his situations and thus of learning. To do less would be to do nothing or to obstruct the learning process. The problem of determining just how much the teacher should enter into the situation where the pupil is confronted by subject matter is the fundamental problem for lesson assignment and for supervised study. In so far as the data at hand can be used, we shall point out where the teacher in the new work of geometry should direct the thinking of pupils and also suggest where this guidance of the teacher is unnecessary and undesirable.

Obviously, if a class is composed of a group of pupils of abilities normally distributed, no one lesson assignment can meet the needs of all pupils save, perhaps, in the beginning of a new topic, and even then uniform assignments as to quantity of work should be avoided. Since one assignment does not suffice for all, it is the office of the teacher to pitch the assignment to the ability of the average pupil, to give the necessary individual help to the below-average pupils, and to give supplementary assignments to the superior pupils.

Power in geometry work does not result from repetition of known processes nor from unsuccessful attempts to solve too difficult problems, but rather from the successful solving of problems just difficult enough to demand on the part of the pupil the exercise of his entire mental equipment. The successful solution of a new problem results in satisfaction which fixes the connections between a given situation and its correct outcome. The ideal situation would therefore be one in which each pupil could be set to work at the task which was just difficult enough for him to accomplish. In order that successful outcomes may be made probable when the pupil meets subject matter similar to that of these exercises, the study shows that in the lesson assignment for each exercise a certain amount of direction on the part of the teacher is both justified and necessary.

In all early lessons in plane geometry, and always in new work, the ideal procedure is that in which in the lesson assignment the members of the class, before leaving the class room, call up theorems evidently involved in the solution. By giving the superior pupils an opportunity to mention the theorems which could be used in the solution provides a means of rewarding them and makes successful solution on the part of the less able pupils more probable. For these pupils will, with this help, do more than they would otherwise have done, and this means progress for them.

The Necessity for an Exact Statement of the Problem. The statements of Exercises 5, 7, 8, 10, 12, and 13 seem on the whole to have been correctly interpreted by pupils and are therefore typical examples of the exact statement of an exercise.

In the first part of Exercise 6 it must have been the purpose of the examiners to measure the knowledge of the theorem: "If two chords intersect within a circle, the product of the segments of one chord equals the product of the segments of the other." Now this general theorem is used by but 46.0 per cent of the pupils. Had the exercise read: "Find first the product of the segments of any chord through this point; then find the length of the shortest chord which can be drawn through this point," the pupils would have been required to use this general theorem in finding the segments of the chord.

In Exercise 8 the problem states that the construction lines must be clearly shown, yet only 88.6 per cent of the boys and 77.5 per cent of the girls show construction lines clearly. These last percentages show that many pupils neglect to consider even obvious elements of the problem. This is a serious fault and should be overcome by painstaking teaching to develop correct habits of work. No statement of a requirement by an examiner could be more clear.

The statements of parts A), B) and C) of Exercise 9 seem to have been clear to the pupils. Part D), however, was avoided by 34.9 per cent of those who selected the exercise as a whole. Of this group only 13.5 per cent seem to comprehend the meaning of this part of the problem. Had the question read, "Find the greatest and least possible lengths of the third side," the knowledge of the same facts would have been tested and the exercise would probably have been more generally understood.

The first part of Exercise 11 was clearly understood. On the second part, however, the pupils failed generally for only 25.5 per cent give evidence of clear thinking. Again only 5.7 per cent of the pupils correctly state the required method. Further, 15.0 per cent of the pupils misunderstood the statement and assumed that the external point was not given or fixed. These then gave correct solutions for the exercise as interpreted. To have made this point more clear would have improved the statement of the exercise. This result could have been accomplished by designating the external point as given or fixed.

In all work with numerical data the computer must determine the degree of accuracy which the problem requires. Two features of the result must be considered: (*a*) the unit in which the result is to be expressed, and (*b*) the number of decimal places to which the computation must be carried. In an examination set to test the knowledge of geometric facts possessed by pupils, uniform responses are more probable if the nature of the result and the degree of accuracy required are stated in the exercise. Classroom practice may well be provided on these two points, but in a set examination, specific direction on these points should be given. In this examination, directions covering these points should have been given for the list as a whole or specifically for Exercises 5, 6, and 9.

Identification of Elements of a Problem. The results of work done by pupils in solving Exercise 5 after a year of study in plane geometry tend to show that the teacher should so direct the presentation of such an exercise in a lesson assignment as:

1. To bring out the fact that the area of a square is given and that the area of a circle is required.

2. To make clear the fact that since the area of the square is given in square inches, it is necessary that the area of the circle be expressed in square inches or square units.

3. To point out or identify all the essential elements of the exercise, area, square, inscribed square, square inches, and area of circle.

4. To lead the pupils in the class room to call up theorems which involve the known and required elements. A varied response on this point will often result in the use of many different methods of solving the exercise—an outcome to be desired. This

recalling of theorems and identification of them as useful affords the best possible kind of review on them.

5. To see that the same elements used in two ways, even where a numerical result is required, should be shown common to both situations—only 1.9 per cent identify the diagonal of the square as the diameter of the circle and take care to point out that fact. Many would doubtless have done this had they been requested to do so, but the percentage given above shows how weak is the habit of critically examining situations and pointing out essential facts in important steps of the process.

Since only 57.5 per cent arrive at the correct result to this relatively simple exercise after a year spent in studying geometry, it is evident that teachers may well spend more class time (*a*) in teaching pupils to identify the given elements of their exercises and the elements of desired outcomes, and (*b*) in calling up theorems which may be employed in using the given data to reach the desired outcomes.

In Exercise 6 only 36.5 per cent of the pupils understand from the reading of the exercise that they are expected first to find the product of the segments of any chord through the given point. The second part of the exercise requires that the pupil find the length of the shortest chord. Clearly, if this exercise were given in a lesson assignment for the first time, the teacher should lead the pupils to see that two distinct outcomes are required and that these outcomes should be reached in the order mentioned.

Wherever possible, teachers should show the necessity of obtaining general outcomes. Here we have a good example of the need as existing even after a year's study, for 43.9 per cent of the pupils find the product of the two halves of the shortest chord drawn through the given point. Many of these pupils may have known that their products were the same as the products of the two parts of any chord through the point, and failed to state the fact. The exercise evidently aims to call out the use of the general relationship, and pupils do not meet this test of their knowledge.

As in Exercise 5, the need for designating the character of the product in Exercise 6—that is, the unit of measurement in it—should be made clear in lesson assignments, for after the usual year of study the work here investigated shows that only

9.2 per cent of the pupils note that the product of these lines should be expressed in square inches.

In their solutions of Exercise 7, pupils unaided seem to fail to identify and use the most obvious outcomes of each element of their exercise. Here they fail to see that they really have the two segments of the shorter side, the given segment of 2 feet, and hence the other segment of (10 — 2) feet or 8 feet. When given the two segments of one side of a triangle (or the equivalent) and required to find the two parts of a second known side made by a line parallel to the base and cutting both of them, only 32.7 per cent of the pupils recall a really important theorem (*A line parallel to the base of a triangle divides the two sides into proportional segments*) and correctly use it in finding the parts of the second side. Now 37.4 per cent of the pupils use the above data or the equivalent in a theorem of far less general application. Apparently the teaching of geometry does not of necessity result in the use of the theorems of widest application.

The work on Exercise 8 shows that 16.7 per cent of the pupils (9.5 per cent of the boys and 22.4 per cent of the girls) fail to note the fact that the circle, and the line outside of it, exist when the exercise is begun and that but one line (or two lines) should be drawn. Such an analysis of an exercise is essential. Pupils should be directed in their reading of exercises to determine precisely what is given and what is to be proved, computed or constructed.

The data show that even on their final papers 13.9 per cent of the pupils drew unnecessary construction lines, lines which here indicated a hit or miss approach to the solution of a relatively simple exercise. A study of these various constructions leads to the conclusion that a rough sketch of the figure showing both the given and the required lines would, if made, so direct the thought of the pupil in reaching his goal that the tendency toward much aimless thinking or random guessing would be avoided.

One pupil from among the 1082 whose papers on this exercise were examined actually drew a rough diagram picturing the given conditions, the circle and the line outside of it, and then by taking a common-sense view of the situation, drew a line which appeared to meet the required conditions (parallel to the given line and tangent to the given circle). Then seeing what was to be done, he set about in another carefully drawn figure to construct a line meeting the two given conditions. Doubtless many other pupils did this work on scratch paper, or perhaps thought it out without setting it down. The practice of sketching the figure (identifying given and required elements) before completing the exact instruction has much to commend it. It is certainly a method commonly used in mechanical drawing and design work. If pupils through lesson assignment and supervised study were trained in this way of identifying the elements of their construction problems, there seems to be ground for belief that a habit might thus be formed which would carry over into the solution of many and varied construction exercises.

From part A of Exercise 9 we again see the necessity of teaching pupils to identify each of the several elements in their exercises. Here 18.8 per cent of the pupils fail to see that the exterior angles of a regular polygon are equal, and that each here equals 1/5 of a right angle or 18 degrees. The fact that 18.8 per cent of the pupils who selected this exercise failed in the first steps of the necessary reasoning process (the identification of elements and the noting of initial outcomes) shows that teachers should direct the work of pupils so as to lead to the formation of habits of correct thinking in fully identifying given elements.

From B of Exercise 9 we find that 22.4 per cent of the pupils fail to solve the exercise. Moreover, many of those who did succeed may have happened upon the correct result by recalling that the relation between central and inscribed angles which intercept the same arc is 2:1. In reading the work of pupils we note the almost total absence (though construction was not required) of a figure which in any way identifies the given and the required elements of the exercise.

From C of Exercise 9 we judge that the lesson assignment should bring out the facts that (1) equilateral triangles are to be compared in area; (2) areas of similar triangles are to each other as squares of corresponding lines; and (3) two corresponding lines can be found from the fact that the side of one triangle equals the altitude of the other. In (2) above, work done by pupils indicates that the fact was known and used by but 17.7

per cent of the pupils. The worth of and means of computing the side of the second equilateral triangle seem to be known to but 17.1 per cent of the pupils who elect to be examined on the exercise. In assigning for the first time a lesson involving an exercise similar to part C, teachers should, on the evidence here presented, point out the essential facts involved in this sort of exercise. As has been said before, and as may be said again, this pointing out of facts and relations which may be useful in new work may be done largely by the superior pupils as their specific contribution to class room work.

From part D of Exercise 9 we find that but 13.5 per cent of the pupils have even an approximately correct concept of what the exercise involves. It is evident that in lesson assignments on similar work teachers should do no less than direct the pupils to see what such an exercise really means—the greatest possible and least possible lengths of the third side of the triangle are required. The responses made by pupils on this exercise are marked more by their dissimilarity than by their likeness. Without specific direction, even after a full year of study, a large percentage of the pupils seem unable to give that response which was evidently expected of them by the exercise as here stated.

From Exercise 10 we see that pupils are quite able to meet a situation where they are required to prove two lines equal (*MB* and *MC*) by means of right triangles if the necessary data are given. It follows, then, that in the assignment of such exercises less and less specific direction is needed.

The work done by pupils in solving the first part of Exercise 11 is so uniformly satisfactory as to lead to the conclusion that after some little experience in solving originals in geometry, the pupil may safely be left to his own resources in an original exercise of this degree of difficulty. Apparently the necessity of the teacher's assistance in identifying the several elements of such situations is obviated by the detailed statement of and specific mention of the several elements in the exercise itself.

The data showing the attempts made by pupils to meet the second requirement—to state a simple method of drawing such a secant—indicate that the pupil, left to himself, will be almost hopelessly lost in the solution, for but 5.7 per cent of the pupils succeed in stating correctly the proper method. Where the attempts at the solution had progressed far enough to give a clue to the method used by the pupil, we find that pupils generally fail to recognize the fact that the given point outside the circle is fixed and that it is from that point that the construction of the required secant is to be begun. Instead they proceed by following directly the lead of the exercise, that is, by drawing a chord and then producing it, its own length, through one of its extremities. The point thus located is taken as the "External" point mentioned in the exercise.

From the facts quoted above, it appears that teachers should, in their lesson assignments of such exercises, bring out the fact that certain points must be considered as determined or fixed by the conditions of the exercise. From such points construction must be begun. Such points are elements in the given situation. Could any work of the teacher be of more value to the pupil than that which helps in the formation of habits of critically examining each situation in order to determine just what are and what are not the given and required elements therein?

In a case such as this, where only 5.7 per cent of the pupils successfully solved the problem (see page 18), it is often worth while to examine the situation to determine, if possible, the real cause for the poor record. Here, though the purpose of the examiner is clear to a mature thinker, yet, since the exercise does not specifically state that the point from which the secant is to start is fixed or given, pupils are evidently in the habit of thinking of a point thus designated as not fixed and proceed on this hypothesis to satisfy the other requirements of the exercise. It follows, then, that pupils need training in the determination of what is given and what is to be proved or constructed when involved situations are presented. Such situations are sure to be encountered, hence training is necessary to insure a correct response.

From the work done by pupils in their attempts to solve Exercise 13, it appears that only 41.7 per cent of the pupils drew a figure which is satisfactory and which represents the lines given, implied, and required. It is evident that a majority of pupils who attempted to solve this exercise began to think of methods to be applied in the solution before identifying the several given elements of it.

Uneconomical Methods. Certain methods revealed by the study show a tendency on the part of the pupil to draw unnecessary construction lines. The construction line, when drawn, provides a new situation which at best more or less directly involves certain given elements. In no case does the drawing of the construction lines make unnecessary the use of the essential elements in the given situation. These lines when drawn may aid the pupil in reaching the desired goal, but in practically every case where construction lines were drawn when not needed, the proof was unnecessarily complicated. Our data show that for Exercise 12 but 15.8 per cent of the boys and 14.2 per cent of the girls succeed in solving the exercise without drawing these unnecessary lines. The value of an ideal of method in attacking even original situations must be evident.

From the above, the conclusion is not warranted that in an exercise involving demonstration, construction lines should be avoided. It does follow, however, that such lines should be drawn only when the situation after careful analysis does not reveal a method of reaching the desired goal. Such lines as are drawn should always provide situations which involve the known and unknown elements, and give promise, thereby, of revealing a fact or relation which was not obvious at the outset.

After having solved an exercise in a direct and seemingly easy manner, it should challenge the ingenuity of the superior pupil who has time at his disposal to devise new and unusual methods of proof. The one who can invent the most methods of proof, not the one who thinks out the most involved method, is the best original thinker. To ignore the obvious and easy methods of work is a mark not of originality but of obstinacy. The really original thinker solves his many problems in the simplest possible way and accomplishes, thereby, more useful work. The ability to analyze a situation so as to see that which is essential to its solution is an ability to be coveted.

Coupling Initial Outcomes with Unused Given Elements. From Exercise 12, as is noted elsewhere, we see the too common tendency on the part of the pupil to complicate the solution by the introduction of unnecessary construction lines. Of course where construction lines are drawn and correctly used, there can be no criticism on the solution, save on the score of economy of time and effort expended. To solve the exercise in a less obvious way may sometimes mark the independent or original thinker. Such methods should be employed only in devising supplementary and unique solutions, and the exercise should first be solved by all through the use of more direct methods.

We find here again the use of many different methods of solving an exercise, each one of which is determined by a combination of two factors—the mind set of the individual and the potency of the several elements in the given exercise. This potency of a given element is shown here by the fact that line *MN*, the perpendicular (see page 22), appears to be parallel to the base of the triangle, and 73.7 per cent of the pupils so prove and so use it. From this point the thinking of many pupils shows an uneconomical tendency, for having used two of the three essential elements of the given exercise, a wise use of the result thus far obtained (the lines are parallel) would be to couple this result with the other essential element (*M* is the midpoint of *AB*) and to note the probable outcome. To have done this in the case of 82.5 per cent of the pupils would have brought the situation up to this point—a certain line is parallel to the base of a triangle and bisects one side of it; then the student who has the necessary information to deal successfully with materials at hand will recall the theorem which has been previously proved, "If a line is parallel to the base of a triangle and bisects one side of it, it bisects the other side of it also." We find, however, that only 40.3 per cent of the pupils who reached the conclusion that the lines are parallel do actually connect this outcome with the third essential element of the situation. Instead, they proceed to draw construction lines and to search for or to make triangles by which they may prove the segments *AN* and *NC* equal (see page 22). Now, since to prove these lines equal, reference must be made in any line of proof to the fact that *M* bisects *AB*, it seems reasonable to conclude that the habit of coupling initial outcomes with the previously unused element or elements is not strong. It is, however, a habit essential to success in the solution of geometrical situations. Since the habits of thought can best be observed by teachers in the first reactions made by the pupils in the lesson assignment, it seems clear that good teaching in geometry does and should encourage in the study lesson the method of thinking described above.

In Exercise 10, the proof of the fact that $AX=MX$ (see page 18), even though relatively simple, seems here quite beyond the ability of about half the pupils; for but 49.7 per cent of the pupils arrive at the result through any line of reasoning, however involved. Here only 8.2 per cent of the pupils reach the conclusion in some one of the ways which does not involve the drawing of the unnecessary construction lines. Again there seems to be a strong tendency for the pupil to draw construction lines forming new triangles before surveying the exercise to see what are the necessary outcomes of his given elements. Only 18 per cent of the pupils see that AD and MP are parallel and recognize P as the mid-point of XD, even though XP is given equal to DP.

Clear thinking and economical methods here, as elsewhere, result only after a critical observation of and grouping of the given elements and their obvious outcomes before supplying additional construction lines. Such thinking does not take place uniformly, as is shown by the results of the study. Intelligent direction on the part of the teacher should bring about better thinking habits.

A Typical Instance of Need for Classroom Drill. Probably no one theorem is more necessary to success in mathematics and the mathematical sciences than is the theorem which expresses the relation between the sides of a right triangle. It follows then that one of the subsidiary aims of the teaching of plane geometry should be to make such use of this relationship as will result in practically perfect response, to those geometric situations or parts of situations for which the knowledge of this fact is essential.

In the finding of the radius in Exercise 5, 7.3 per cent of the boys and 9.2 per cent of the girls failed to make a correct use of the right triangle relationships. In Exercise 6, 12.1 per cent of the boys and 9.0 per cent of the girls failed in this respect. In part C of Exercise 9, pupils are required to use or find the relation between the altitude and base of an equilateral triangle. This particular exercise (Exercise 9) is selected by only 19.8 per cent of the boys and 16.2 per cent of the girls, yet of those who do select it only 17.9 per cent of the boys and 16.5 per cent of the girls recall the relation of altitude to side or find this relation through the use of the right triangle theorem. Even though the exercise is selected by a small percentage of pupils, and those presumably the more able, yet this group of pupils has by no means mastered the use of the theorem. We must bear in mind, too, that the papers read are those of pupils who have passed the examination as a whole. It would seem reasonable to assume therefore that had the papers of those who failed to pass the examination been examined also, as large a percentage at least would be found wanting in their knowledge of this important theorem.

Since after a year of study on geometry all pupils or practically all pupils may not be depended upon for correct solutions of exercises involving simple applications of this theorem, there seems to be need for more applications of this theorem in such a course. In the applications of this theorem, pupils should be drilled until a perfect response is assured.

Interpretation of Results Obtained from the Solutions. When an exercise involves units of length or area, the best practice demands that the result should be stated in terms of proper units.

In Exercise 5, the area of the inscribed square is expressed in square inches. Hence the area of the circle should be expressed in square inches. Records made on this point show that only 48.2 per cent of the boys and 39.6 per cent of the girls do so express it. Through carelessness 1.8 per cent of the boys and 2.1 per cent of the girls express the area in inches. It is evident that the element "square inches" as given in the exercise was not of sufficient potency to bring forth the general response—the area of the circle must be expressed in square units.

In Exercise 6, the radius is given and the distance from the center to the given point. Now the careful student will observe that by producing the radius through the given point so as to form a diameter, a chord is drawn whose parts are known. From these data, one then can find the product of the segments of any chord through the point by use of the theorem: "If two chords are drawn through a point within a circle, the product of the segments of one chord equals the product of the segments of the other." To obtain this product is the first requirement of the problem; yet 51.2 per cent of the boys and 73.1 per cent of the girls show that the theorem quoted above is not potent, for they proceed to find the length of the shortest chord first and then to multiply its halves together.

This method of thinking is perfectly correct, yet it does not show a habit of thinking clearly through a situation, and of satisfying each integral part of the exercise as data are provided. It does reveal a tendency to obtain the desired result through the use of a special case. Such thinking is superficial and should prove unsatisfying to the pupil whose thinking is directed by a teacher. No result should be accepted by the pupil as final until it is seen to satisfy the evident purpose of the exercise.

In the same exercise the length of the shortest chord is required. However, only 26.9 per cent of the boys and 27.8 per cent of the girls take care to express the length of the chord in linear units.

In Exercise 7 certain parts of the triangle are expressed in terms of feet and the length of certain other lines is required. However, only 28.9 per cent of the boys and 28.0 per cent of the girls do express the length of the required parts in terms of the linear unit, feet. Again, 4.8 per cent of the boys and 3.3 per cent of the girls express one of the required parts in terms of feet, but neglect to express other required lengths in terms of a linear unit.

In Exercise 9, it is seen that pupils who elected to solve this exercise habitually think of an angle or an arc as expressed in degrees or minutes, for practically no pupil who arrived at a result for the second part of this exercise failed to express that result in degrees. In part D of this exercise, pupils should state their results in terms of feet. The strength of this habit is not revealed here, for only 61.7 per cent of the boys and 75.3 per cent of the girls who select Exercise 9 attempt part D of it. Among those who select this part (19.3 per cent of the boys and 6.6 per cent of the girls) and react intelligently to it, there is seen a strong tendency to express the result in linear units. Apparently the more careless students have either not chosen the exercise at all, or having chosen it, have left no record of their responses. The fact that the exercise is worded in such a way as to require careful examination may itself bring to consciousness the unit or feet element in the situation.

For Exercise 5, records show that only 48.2 per cent of the boys and 39.6 per cent of the girls express their results in terms of proper units, and since 61.6 per cent of the boys and 53.9 per cent of the girls were given perfect scores (see pages 46 and 106), it is evident that teachers and examiners do not consider this expressing results in terms of proper units as necessary to a correct and adequate solution of such exercises. Comparing the per cents of perfect scores in the several exercises with the per cents of results stated in terms of proper units, we have the following:

Exercise	Per Cent of Results Properly Stated		Per Cent of Results Scored Perfect	
	Boys	*Girls*	*Boys*	*Girls*
No. 5	48.2	39.6	61.5	53.9
No. 6 1st part	7.4	11.0	44.3	34.3
No. 6 2d part	26.9	27.8	44.7	34.7
No. 7	28.9	28.0	61.2	49.2

The above percentages for those whose results were stated properly would have been somewhat altered had the papers written by pupils not passing the subject been examined. It is hardly probable, however, that a larger percentage of such pupils would have stated their results in terms of proper units than of those who did pass the examination.

From the inspection of the above data it seems clear that the teaching of geometry does not automatically result in the development of habits of interpreting results and expressing them in proper units of measurement. Moreover, since teachers and examiners mark papers correct when proper methods of solution have been used, even though results are not adequately expressed, it seems clear that in the teaching of geometry not enough emphasis is placed upon this essential feature.

The materials treated in a modern text offer to the teacher of geometry ample opportunity to assist the pupil in the formation of correct habits of expressing results in terms of proper units. This opportunity should not be neglected, for in the world of practical affairs a result obtained from a solution is meaningless unless the character of the result is properly designated.

From among the 917 (83.6 per cent of 1,058) pupils who select Exercise 11 and satisfy the first part of it, we find that the required statement is given by 2.9 per cent and that but 29 pupils note that it is necessary to state the conditions under which the construction is possible. Here the pupil should have stated that the distance from the external point to the nearest point on the circumference of the circle must not be greater than the diameter of the circle. A habit of dealing with the limiting conditions of all situations involving construction is much needed. It is so important that it should be often strengthened through use in lesson assignments. Pupils readily copy methods employed by the teacher. The exercise here considered implicitly demands that the statement of limiting solution be made even though the requirement is not specifically stated. The habits of teachers and examiners in this matter are shown by the fact that though six pupils made note of limiting conditions, eighty-seven pupils receive practically perfect (12 or 12½) scores. One is led, therefore, to raise the question: Do teachers sufficiently appreciate the necessity of teaching pupils to consider the limiting conditions in construction exercises?

Conclusions. From the data considered in this chapter the evidence is conclusive that teachers should recognize the individual differences of pupils. In instruction based on recognition of these differences the teacher must in lesson assignments prepare each pupil for probable success in dealing with the exercises assigned for home study. To this end, exercises should be clearly stated and the minimum though adequate direction should be given to the several pupils that fundamental theorems may be perfectly learned, that the given and required elements of the original exercises may be identified, and that the direct outcomes of each of the given elements may be pointed out before the pupil leaves the classroom. At times there should be teacher direction and class participation in pointing out such groupings of known elements and initial outcomes as give promise of being useful in reaching the desired goal. Class participation in the solution of original exercises is needed that habits of economical thinking may be developed, such as the full identification of all given elements, the drawing of construction lines only when needed, the coupling of initial outcomes with the unused given elements, and in the testing of results. This class instruction should always be supplemented by necessary individual instruction. It is evident also that pupils even after a year of study of geometry have not formed habits of examining their results to determine whether or not the evident and implied requirements of the exercise have been satisfied and the limiting conditions considered. Such habits should be formed early in the process of teaching pupils how to study and should be strengthened through continued exercise in class solutions of new and difficult originals.

CHAPTER VI

RESUMÉ OF THE STUDY

Data Used and Methods Employed. The abilities shown by second year high school pupils in solving geometric originals have been investigated in this study. The materials used were 2800 plane geometry examination papers written by New York state high school pupils in their Regents examination of June, 1918. The ninety schools from which the papers were selected were among the better schools of the state and since in the study access was not had to the papers of pupils who failed in the examination as a whole, the findings are based upon work done by the better than average pupils of the state.

In the study, papers were drawn from the several schools in such a way as to afford a fair sampling of the work done by pupils in each of the several high schools and also to provide for investigation and analysis approximately five hundred papers for each sex group. The analysis of this sizable sampling affords reliable data on interests and abilities of high school boys and girls upon which to base conclusions for improved methods of teaching plane geometry.

Findings of the Study and Recommendations. The findings of the study give tentative answers to the following questions:

1. What differences in difficulty are inherent in the originals of a given examination?

2. What differences in interests are shown by the selection of certain exercises rather than certain others, and what differences in difficulty are experienced by boys and girls in solving construction, numerical, and proof originals?

3. Is there a general method which should be consciously followed in the solution of geometric originals in order that correct solutions may be successfully and economically obtained?

4. What suggestions grow out of the study which bear on the improvement of the teaching of geometry?

Since on taking the examination pupils were directed to solve eight of the thirteen theorems and exercises, the percentages of pupils selecting the several exercises serve as a test of the appeal made by each type of exercise. The study shows that pupils prefer to solve construction exercises rather than exercises involving numerical or demonstrative work and that solutions involving proof or demonstration are on the whole least often chosen. The mean scores made show that pupils experience least difficulty in solving the construction exercise and that the greatest difficulty is experienced in solutions involving demonstration.

Through the use of a composite measure based on the frequency with which the exercise is chosen and on the mean scores made by those who solved the several exercises, it is found that in this examination, which is doubtless more carefully worked out than most school examinations, two of the exercises are respectively nine and fourteen times as difficult as other exercises in the set. Even though great differences in difficulty must have been apparent, correct solutions of the several exercises were scored as of equal worth.

The most pronounced preference for one type of exercise rather than another is that shown by boys in selecting the construction exercise. Here the correlation (tetrachoric r) between being a boy and selecting the exercise was $r = +0.31$. This preference is probably due to the strength of the manipulation-experimentation interest in boys. Girls also select this exercise more often (with one exception) than an exercise involving numerical solution or proof. That there is a real sex difference in the choice of the construction exercise is shown by the fact that were there really no sex difference, a difference in the percentages of times this exercise was chosen by boys and girls would be found as a matter of chance to be as great as that which was observed only once in 559,000 times.

The several measures of ability used all show that the male sex slightly excels the female in solving original exercises. There is a positive though low correlation (the bi-serial r) between being a boy and excelling the median of the total group in solving each original. The correlation coefficients range from $+0.07$ to $+0.38$ with an average of $+0.21$. The study shows however that the differences between the sexes in abilities here observed

(61 per cent of the boys equal or excel the median girl) are far outweighed in importance by the variability within each sex group. On the score of geometric abilities, therefore, no evidence is here found on which to base a claim for segregating the sexes for the study of plane geometry.

The study shows that the amount of overlapping of the sex groups in abilities to solve numerical exercises is approximately equal to the overlapping shown in abilities to solve exercises involving demonstration. Again the range in boy to girl variabilities is as great on numerical exercises as on exercises involving demonstration. There seems therefore to be no good reasons for so changing the content of the course of study as to assign numerical exercises to one sex group and exercises involving demonstration to the other.

A study of the work of pupils in solving geometric originals suggests a formulation of the steps of the thought processes which is here given:

I. Identification of
   - A. All given elements (facts or relations).
   - B. Elements of the desired outcome (facts or relations to be proved).

II. Search for
   - A. A direct outcome of each given element considered alone and apparent implications of the desired outcome.
   - B. Situations (combinations of elements) which do or would involve known elements, or better, known elements and the desired outcome.

III. Selection of a certain situation because it involves most known elements or potent elements which give promise of being useful in reaching or leading toward the desired goal.

IV. Utilization of a situation and its outcomes as a new situation, etc., leading toward the desired outcome. This step includes or may include:
   - A. The testing of a selected situation and its outcomes in an effort to reach the desired outcome, or
   - B. The testing and rejecting of a selected situation and its outcomes as of no worth in reaching the desired outcome, or

C. The testing and rejecting as in B above followed by the drawing of construction lines which provide a new situation to be tested as in A or B above.

V. Verification of conclusions in:

A. Problems involving numerical data through interpretation of and checking of results.

B. Problems involving construction through the consideration of limiting conditions and proof that the construction is correct.

C. Problems involving proof through syllogistic organization.

Some additional suggestions which grow out of the study are:

1. Wide differences in abilities do exist even among pupils who pass the examination as a whole. These differences among pupils should be recognized in lesson assignments and teaching that success for each pupil may be made probable.

2. More care than is commonly exercised should be placed on the exact statement of the several exercises of an examination.

3. In the lesson assignment on new and really difficult originals the teacher should direct the class in identifying the several given and required elements of the problem.

4. In lesson assignments, pupils should be made conscious of the need for coupling initial outcomes with the previously unused given elements found in the problem situation presented.

5. Continued classroom drill on fundamental theorems is necessary for even the right triangle relation is not sufficiently well learned to guarantee its correct use where needed.

6. Pupils should be taught to interpret and to check the results obtained from each solution. That such habits of dealing with results are not formed is obvious from the study.

APPENDIX

COMPUTATION OF MEASURES

Use Made of Measures of Central Tendency and Variability

In the pages which follow there are shown the data upon which the computation of the several measures of central tendency and variability, and their probable errors, are based. Since the median and the quartile are measures easily computed and not infrequently used, the method of computing these measures is not here given. The distribution of the scores for both boys and girls on each exercise is given, and in connection with these distributions for Exercise 5 the actual computations of the mean and sigma, σ, are included. For Exercises 6 to 13, only the results for these measures are indicated.

It is often less difficult to compute the mean score and the σ for a combined group from data showing these measures for two separate groups than to combine the distributions and to compute these measures for the combined distribution. The method of computing the means for the combined groups on each exercise is given on page 107 and the method of computing the standard deviations on page 108. The probable errors of these measures for the combined groups, computed as in the manner mentioned above, are recorded in connection with the P.E.'s for the separate groups (see page 113).

TABLE XVI

COMPUTATION OF MEASURES OF CENTRAL TENDENCY AND VARIABILITY OF THE SCORES MADE BY BOYS ON EXERCISE 5

Scores (X's) 3 = 2.75–3.25 and 12.5 = 12.25–12.5	Frequency of Scores (f)	Frequency × Score $f \times X$	Frequency × Square of Score $f \times X^2$
0.0	18	0.0	0.00
0.5	0	0.0	0.00
1.0	6	6.0	6.00
1.5	0	0.0	0.00
2.0	15	30.0	60.00
2.5	0	0.0	0.00
3.0	7	21.0	63.00
3.5	0	0.0	0.00
4.0	15	60.0	240.00
4.5	0	0.0	0.00
5.0	5	25.0	125.00
5.5	0	0.0	0.00
6.0	34	204.0	1224.00
6.5	8	52.0	338.00
7.0	5	35.0	245.00
7.5	27	202.5	1518.75
8.0	10	80.0	640.00
8.5	5	42.5	361.25
9.0	8	72.0	648.00
9.5	2	19.0	180.05
10.0	5	210.0	2100.00
10.5	21	52.5	551.25
11.0	16	176.0	1936.00
11.5	3	34.5	396.75
12.0	33	396.0	4752.00
12.5	303	3787.5	47343.75
	$N = 546$	Sum or $(\Sigma X) = 5505.5$	$\Sigma(f \times X^2) = 627298.80$

Mean, or average $= \frac{\Sigma(f \times X)}{N} = \frac{5505.5}{546} = 10.08.$

σ, or Standard Deviation of the Distribution $= \sqrt{\frac{\Sigma(f \times X^2)}{N} - M^2} = \sqrt{\frac{627298.80}{546} - 101.60} = 3.64.$

Median = 12.29, the score below which and above which just 50 per cent of the cases lie.

Q_1, or 25-percentile = 7.69, the score below which 25 per cent of the cases lie.

Q_3, or 75-percentile = 12.39, the score below which 75 per cent of the cases lie.

Q, or semi-interquartile-range $= \frac{Q_3 - Q_1}{2} = \frac{12.39 - 7.69}{2} = 2.35$ or 2.4 (½ range of middle 50 per cent).

The above measures of central tendency and variability for the scores made by boys and girls on the several exercises are given in Table XVII.

TABLE XVII. DISTRIBUTIONS OF SCORES ON THE SEVERAL EXERCISES

Exercise: Score 1 = 0.75 to 1.25; 12.5 = 12.25 to 12.5	No. 5 Freq. Boys	Girls	No. 6 Freq. B.	G.	No. 7 Freq. B.	G.	No. 8 Freq. B.	G.	No. 9 Freq. B.	G.	No. 10 Freq. B.	G.	No. 11 Freq. B.	G.	No. 12 Freq. B.	G.	No. 13 Freq. B.	G.
0	18	15	21	32	11	33	21	46	10	15	4	8	31	39	16	41	30	69
0.5	0	0	0	0	0	0	0	0	0	0	0	0	0	1	0	0	5	4
1.0	6	1	4	1	0	1	0	0	5	4	2	2	0	3	1	3	2	10
1.5	0	0	0	0	0	0	0	0	2	0	0	0	1	0	2	1	1	3
2.0	15	20	5	7	0	3	2	1	3	1	4	3	2	14	7	10	18	35
2.5	0	3	0	4	0	1	0	1	0	1	0	0	5	2	3	4	1	13
3.0	7	26	7	9	2	7	6	41	37	59	0	4	9	25	7	13	55	129
3.5	0	1	2	2	1	0	0	1	2	14	0	0	1	0	0	0	0	1
4.0	15	14	6	16	15	25	6	23	19	11	9	7	10	13	7	13	7	12
4.5	0	2	4	4	3	2	0	0	10	6	0	1	0	3	2	2	0	3
5.0	5	15	6	11	14	13	1	0	4	8	4	13	6	9	0	3	6	6
5.5	0	1	0	2	2	4	0	1	0	3	0	2	1	1	1	0	0	5
6.0	34	35	45	32	44	38	33	34	70	51	18	59	38	59	9	6	1	3
6.5	8	5	43	54	32	59	43	100	8	24	78	151	8	12	1	6	0	1
7.0	5	5	17	23	1	26	1	0	11	17	23	19	32	24	1	3	0	0
7.5	27	39	8	9	1	4	1	2	10	4	23	15	3	7	1	1	0	1
8.0	10	8	20	16	9	10	2	6	19	7	15	11	23	38	9	10	2	3
8.5	5	8	7	7	3	7	0	2	3	0	6	14	19	4	3	3	0	1
9.0	8	9	6	7	1	6	0	0	16	5	10	15	118	207	5	8	0	1
9.5	2	11	7	9	1	7	1	1	13	7	6	1	35	12	0	4	1	0
10.0	21	30	55	53	22	26	0	0	5	1	20	17	61	44	21	18	1	0
10.5	5	6	11	29	15	30	0	0	0	1	10	12	9	2	16	13	3	1
11.0	16	14	7	12	12	12	0	2	4	0	11	19	28	4	19	19	2	0
11.5	3	12	6	7	4	4	1	1	0	0	10	20	4	3	18	23	3	1
12.0	33	33	68	20	114	34	37	12	1	0	43	48	16	9	94	85	4	0
12.5	303	294	163	161	191	274	346	305	4	4	168	111	46	16	304	281	20	7
Med. Boys	12.3		10.2		12.0		12.3		6.0		10.7		9.0		12.3		3.0	
Med. Girls	12.1		9.9		11.2		12.3		5.6		7.5		8.8		12.2		2.8	
Q. Boys	2.4		2.9		2.8		0.3		1.9		3.0		1.6		0.5		1.3	
Q. Girls	2.5		3.0		2.9		3.1		1.7		2.8		1.6		1.2		1.1	
Mean Boys	10.1		9.4		10.2		10.7		5.9		9.7		8.3		11.0		4.3	
Mean Girls	9.7		8.8		9.4		8.9		5.0		8.6		7.3		10.2		2.6	
S.D. Boys	3.6		3.5		3.2		3.4		2.7		3.0		3.1		3.1		4.2	
S.D. Girls	3.7		3.7		3.7		4.3		2.5		3.0		3.1		4.0		2.4	

See page 114 for values of the above measures to 2-place accuracy.

Computation of the Means of the Combined Distributions

To find the mean of the combined group (M_c) from the mean of the distributions for the boys' group (M_b) and of the girls' group (M_g):

Let N_b = number in boys' group, and N_g = number in girls' group, and $N_c = N_b + N_g$.

Now $M_c = \frac{M_b \cdot N_b + M_g \cdot N_g}{N_c}$

Then for Exercise 5, we have,

$$M_c = \frac{(10.08)\ (546) + (9.70)\ (607)}{1153} = 9.88$$

The means for the combined groups as computed in the case of Exercise 5 above are given in Table XVIII.

Probable Errors of the Medians

To find P.E.$_{\text{Median}}$ of boys' distribution on Exercise 5:

P.E.$_{\text{Median}} = 0.6745\ \frac{\sqrt{N}}{2} \cdot \frac{i}{f}$ where N = the population, i = the total interval used (after smoothing curve in vicinity of median), and f = frequency of cases in that interval.

Here $N = 546$, $i = 0.75$ or (12.50–11.75) which is the interval affected by the smoothing process, and $f = 336$ or $(33 + 303)$ the cases in the interval considered.

$$\text{P.E.}_{\text{Median}_b} = 0.6745\,\frac{\sqrt{546}}{2} \times \frac{0.75}{336} = 0.018.$$

$$\text{P.E.}_{\text{Median}_g} = 0.6745\,\frac{\sqrt{607}}{2} \times \frac{0.75}{327} = 0.019.$$

$$\text{P.E.}_{\text{Median}_c} = 0.6745\,\frac{\sqrt{1153}}{2} \times \frac{0.75}{663} = 0.01.$$

The medians of the several distributions along with the probable errors as computed in the case of Exercise 5 above are given in Table XVIII.

STANDARD DEVIATIONS OF THE COMBINED DISTRIBUTIONS

To compute the standard deviation of the combined boy-girl distribution from the known frequencies and measures of central tendency and variability for the boys' group and girls' group, use is made of the following formula

$$\sigma_c = \sqrt{\frac{N_b(\sigma_b^2 + M_b^2) + N_g(\sigma_g^2 + M_g^2)}{N_c} - M^2_c}$$

in which

N_b, N_g, and N_c represent the numbers in the several groups,

M_b, M_g, and M_c represent the means of the several groups,

and

σ_b, σ_g, and σ_c represent the standard deviations of the several groups.

Then for Exercise 5, we have

$$\sigma_c = \sqrt{\frac{546(3.64^2 + 10.08^2) + 607\,(3.67^2 + 9.70^2)}{1153} - 9.88^2}$$

$$= \sqrt{\frac{627108.30 + 652883.13}{1153} - 97.616}$$

$$= \sqrt{13.398}$$

$$\sigma_c = 3.66.$$

The standard deviations of the combined distributions for the several exercises, computed as in the case of Exercise 5 above, are given in Table XVIII.

Probable Errors of the Means

To find the Probable Error of the Mean for each distribution: Let M_b, M_g, and M_c designate the Means of the boys', girls', and combined groups respectively.

Since $\text{P. E.}_{\text{Mean}} = 0.6745 \frac{\sigma}{\sqrt{N}}$, then for the distribution of Exercise 5, we have

$$\text{P.E.}_{M_b} = 0.6745 \frac{3.64}{\sqrt{546}} = 0.11.$$

$$\text{P.E.}_{M_g} = 0.6745 \frac{3.67}{\sqrt{607}} = 0.10.$$

$$\text{P.E.}_{M_c} = 0.6745 \frac{3.66}{\sqrt{1153}} = 0.07.$$

The mean scores for the several distributions along with their probable errors, computed as in the case of Exercise 5 above, are given in Table XVIII.

Probable Errors of the Standard Deviations

To find the Probable Error for the standard deviation (σ) of a distribution, use $\text{P. E.}_{\sigma\ \text{distribution}} = 0.6745 \frac{\sigma}{\sqrt{2N}}$, in which $\sigma = \sigma_{\text{distribution}}$ and N = number of cases included.

Then for Exercise 5, we have

$$\text{P.E.}_{\sigma_b} = 0.6745 \frac{3.64}{\sqrt{1092}} = 0.08.$$

$$\text{P.E.}_{\sigma_g} = 0.6745 \frac{3.67}{\sqrt{1214}} = 0.07.$$

$$\text{P.E.}_{\sigma_c} = 0.6745 \frac{3.66}{\sqrt{2306}} = 0.05.$$

The standard deviations of the several distributions along with their probable errors, computed as in the case of Exercise 5 above, are given in Table XVIII.

PROBABLE ERRORS OF THE SEMI-INTER-QUARTILE RANGES

(*a*) To find the P.E. of the semi-interquartile range (Q), for the boys' distribution on Exercise 5, use Kelley's formula:

$$\sigma_{(uq-lq)_b} = \sqrt{n\frac{3}{16}\cdot\frac{i^2}{f^2} + n\frac{3}{16}\cdot\frac{I^2}{F^2} - \frac{2}{3}n\frac{16}{3}\cdot\frac{f}{i}\cdot\frac{F}{I}}$$

in which uq = upper quartile, lq = lower quartile, n = number in the distribution, i = interval used in smoothing curve at lower quartile, I = interval used in smoothing curve at upper quartile, f = frequency in the interval i, and F = frequency in the interval I.

$$\sigma_{uq} = \sqrt{n\frac{3}{16}\cdot\frac{i}{f}} \quad \text{and} \quad \sigma_{lq} = \sqrt{n\frac{3}{16}\cdot\frac{I}{F}}.$$

Then

$$\sigma_{(uq-lq)_b} = \sigma_{Q_b} = \sqrt{\frac{3n}{16}\cdot\frac{i^2}{f^2} + \frac{3n}{16}\cdot\frac{I^2}{F^2} - \frac{3n}{16}\cdot\frac{2}{3}\cdot\frac{i}{f}\cdot\frac{I}{F}}$$

$$= \sqrt{\frac{3n}{16}\left(\frac{i^2}{f^2} + \frac{I^2}{F^2} - \frac{2}{3}\cdot\frac{i}{f}\cdot\frac{I}{F}\right)}$$

In the boys' distribution for Exercise 5,

$n = 546.0.$
$i = 1.5$ (3 steps used in smoothing the curve in the vicinity of lower quartile).
$f = 42.0$ (frequency in above 3 steps).
$I = 0.25$ (½ step used near upper quartile).
$F = 303.0$ (frequency in above ½ step).

Since

P.E. $= 0.6745\sigma$,

$$\text{P.E.}_{Q_b} = \frac{0.6745}{4}\sqrt{3(546)\left(\frac{1.5^2}{42^2} + \frac{0.25^2}{303^2} - \frac{2}{3}\cdot\frac{1.5}{42}\cdot\frac{0.25}{303}\right)} = 0.24$$

Then

$Q_b = 2.40 \pm 0.24.$

(*b*) Then

$$\text{P.E.}_{Q_g} = \frac{0.6745}{4}\sqrt{3\ (607)\left(\frac{1.5^2}{52^2} + \frac{0.25^2}{294^2} - \frac{2}{3}\cdot\frac{1.5}{52}\cdot\frac{0.25}{294}\right)} = 0.19$$

Then

$Q_g = 2.53 \pm 0.19.$

Probable Errors of the Bi-serial r's

A method of computing the Probable Error of the bi-serial r has not been worked out. The best assumption one can make is that it lies between the P.E of the r of the product-moment correlation coefficient and that of the tetrachoric r. Since the groups considered here are of approximately equal size, we shall use $\text{P.E.}_{r_{b.s.}} = \dfrac{\text{P.E.}_r + \text{P.E.}_{r_t}}{2}$ to find the probable error of the bi-serial r ($\text{P.E.}_{r_{b.s.}}$) for the correlation coefficient for each of the Exercises 5 to 13 inclusive.

For Exercise 5,

$$\text{P.E.}_r \quad (\text{for } r = -0.073 \text{ and } n = 1153) = \frac{0.6745 \times 0.99467}{33.9558} = 0.0198.$$

$$\text{P.E.}_{r_t} \quad (\text{for } r_t = -0.073 \text{ and } n = 1153) = \qquad = 0.0311.$$

$$\text{P. E.}_{r_{b.s.}} = (0.0198 + 0.0311) \div 2 \qquad = 0.03.$$

The bi-serial r's for the scores made in the several exercises by boys and girls with the probable errors, computed as in the case of Exercise 5 above, are given in Table XIX.

Measures of Variability Within Groups

In connection with a measure of central tendency of a group in any trait, we should also consider the variability within that group. When the median is used as a measure of central tendency, the quartile deviation is used as a measure of variability.

When the mean is used as a measure of central tendency, the sigma (σ) of the distribution is used here as the measure of variability. A standard deviation or sigma (σ) indicates the distance which must be laid off (half to the right and half to the left of the mean) so as to include 68.26 per cent of the cases.

It should be noted that the quartile (or Q) indicates the semi-interquartile range and is therefore a measure of the variability of the group. To find Q we must recall that 1 P. E. indicates the distance on the base line of the surface of frequency which must be laid off above and below the median in a normal distribution so that the area above this base line will include 50 per cent of the cases. Thus Q equals one half the range from —1 P. E. to +1 P. E.

The relative value of σ and Q as measures of variability as applied to the distributions for Exercise 5 of this list is below.

Comparison of the Quartile Deviation (Q) and the Standard Deviation (σ) as Measures of Variability Within Groups

That we may compare Q with σ with respect to reliability as measures of variability, we must relate the probable errors of each with the measures themselves. The values for Q and σ for the distributions in Exercise 5 are shown on page 113. The computations of the probable errors of the Q's for the boys' and girls' distributions on Exercise 5 are made through the use of Kelley's formula on page 110. The computations of the probable errors of σ for Exercise 5 are given on page 109. From the data referred to above we have:

For the boys' distribution on Exercise 5

$$Q = 2.40 \text{ and } P.E._Q = 0.244$$
$$\sigma = 3.46 \text{ and } P.E._Q = 0.074.$$

Now we have

$$\frac{P.E._Q}{\sigma} < \frac{P.E._Q}{Q} \quad \text{for} \quad \frac{0.074}{3.64} < \frac{0.244}{2.40}$$

Therefore σ is a better measure of variability for the boys' distribution on Exercise 5 for it is less variable as a measure.

For the girls' distribution on Exercise 5

$$Q = 2.53 \text{ and } P.E._Q = 0.19 \text{ and}$$
$$\sigma = 2.845 \text{ and } P.E._Q = 0.055.$$

Again

$$\frac{P.E._Q}{Q} < \frac{P.E._Q}{\sigma} \quad \text{for} \quad \frac{0.055}{2.85} < \frac{0.19}{2.53}$$

Again it is noted that σ is the more reliable measure of variability.

From what has preceded, we judge σ to be the more reliable measure of variability within our groups for the trait in question. We have therefore used the mean as the measure of central tendency and σ as the measure of variability in scores on these original exercises in geometry.

TABLE XVIII

TABLES OF MEASURES OF ABILITY AND VARIABILITY

Exercise No.	Sex	Median Score Med. ± P.E.$_{Med.}$	Mean Score Mean ± P.E.$_{Mean}$	Semi-inter-quar. Range Q ± P.E.$_Q$	Standard Deviation S.D. ± P.E.$_\sigma$	Coef. of Variability S.D. ÷ Mean	$V_B \div V_G$
5	B	12.29 ± 0.02	10.08 ± 0.11	2.40 ± 0.24	3.64 ± 0.08	0.36	0.95
	G	12.10 ± 0.02	9.70 ± 0.10	2.53 ± 0.19	3.67 ± 0.07	0.38	
	(B & G)	12.26 ± 0.01	9.88 ± 0.07	2.44	3.66 ± 0.05	0.37	
6	B	10.21 ± 0.16	9.36 ± 0.10	2.85	3.49 ± 0.07	0.37	0.88
	G	9.92 ± 0.13	8.76 ± 0.11	2.97	3.72 ± 0.08	0.42	
	(B & G)	10.07 ± 0.11	9.06 ± 0.08	2.92	3.61 ± 0.06	0.40	
7	B	11.99 ± 0.11	10.20 ± 0.10	2.79	3.18 ± 0.07	0.31	0.79
	G	11.21 ± 0.20	9.42 ± 0.10	2.93	3.71 ± 0.07	0.39	
	(B & G)	11.92 ± 0.02	9.77 ± 0.07	2.88	3.49 ± 0.05	0.36	
8	B	12.32 ± 0.02	10.68 ± 0.10	0.28	3.43 ± 0.07	0.32	0.67
	G	12.26 ± 0.02	8.92 ± 0.12	3.09	4.30 ± 0.08	0.48	
	(B & G)	12.29 ± 0.01	9.74 ± 0.08	2.98	4.01 ± 0.06	0.41	
9	B	6.01 ± 0.10	5.85 ± 0.11	1.94	2.68 ± 0.08	0.46	0.92
	G	5.58 ± 0.10	4.93 ± 0.11	1.66	2.49 ± 0.08	0.50	
	(B & G)	5.88 ± 0.07	5.43 ± 0.08	1.81	2.62 ± 0.05	0.48	
10	B	10.75 ± 0.29	9.68 ± 0.09	2.97	2.97 ± 0.06	0.31	0.89
	G	7.48 ± 0.26	8.62 ± 0.09	2.79	3.03 ± 0.06	0.35	
	(B & G)	9.01 ± 0.27	9.16 ± 0.06	2.88	3.06 ± 0.05	0.34	
11	B	9.02 ± 0.07	8.33 ± 0.09	1.62	3.10 ± 0.06	0.37	0.88
	G	8.80 ± 0.06	7.34 ± 0.09	1.57	3.07 ± 0.06	0.42	
	(B & G)	8.88 ± 0.08	7.81 ± 0.06	1.65	3.13 ± 0.04	0.40	
12	B	12.27 ± 0.02	11.02 ± 0.09	0.49	3.06 ± 0.06	0.28	0.72
	G	12.23 ± 0.02	10.16 ± 0.11	1.15	3.96 ± 0.08	0.39	
	(B & G)	12.26 ± 0.16	10.58 ± 0.07	0.76	3.57 ± 0.05	0.34	
13	B	2.97 ± 0.12	4.29 ± 0.22	1.30	4.22 ± 0.16	0.98	1.09
	G	2.83 ± 0.07	2.63 ± 0.09	1.09	2.36 ± 0.06	0.90	
	(B & G)	2.87 ± 0.06	3.20 ± 0.10	1.03	3.23 ± 0.07	1.01	

TABLE XIX

SUMMARY OF MEASURES ON GEOMETRIC INTERESTS AND ACHIEVEMENTS OF BOYS AND GIRLS

The examination contained thirteen parts, four theorems and nine originals, of which any eight might be selected. Of the selections made by the pupils, only the originals were examined in this study.

Exercise Number and Type of Exercise.	Sex	Total of papers referred to in the study.	Per cent of times selected by pupil:	The probability that a random sampling would show as great or greater difference in frequency of selection is:	Correlation in selections: (Tetrachoric r)	Per cent making perfect score:	Mean score: (Max. score 12.5)	Standard Deviation.	Per cent of boys excelling median girl.	Correlation in achievement: (Bi-serial r)	Coefficient of variability: (S.D. ÷ Mean)	Ratio of boy to girl variability.
(A)	(B)	(C)	(D)	(E)	(F)	(G)	(H)	(I)	(J)	(K)	(L)	(M)
5	Boys	792	68.89	1: 1	0.00 ± 0.03	B. 62	10.1	3.6	57	0.07 ± 0.03	0.36	0.95
(Numerical)	Girls	881	68.89			G. 54	9.7	3.7			0.38	
6	B.	1030	50.0	1: 6.2	+ 0.08 ± 0.02	B. 45	9.4	3.5	56	0.17 ± 0.03	0.37	0.88
(Numerical)	G.	1167	45.6			G. 34	8.8	3.7			0.43	
7	B.	621	80.2	1: 3	− 0.09 ± 0.03	B. 61	10.2	3.2	62	0.14 ± 0.03	0.31	0.79
(Numerical)	G.	745	84.0			G. 49	9.4	3.7			0.40	
8	B.	540	92.8	1: 559,000	+ 0.32 ± 0.04	B. 76	10.7	3.4	66	0.38 ± 0.02	0.32	0.67
(Construction)	G.	703	82.4			G. 55	8.9	4.3			0.48	
9	B.	1289	18.9	1: 9.4	− 0.09 ± 0.02	B. 2	5.9	2.6	64	0.33 ± 0.03	0.44	0.92
(Numerical)	G.	1504	16.2			G. 2	4.7	2.3			0.51	
10	B.	842	55.1	1: 2.5	− 0.07 ± 0.02	B. 45	9.7	3.0	67	0.22 ± 0.03	0.31	0.89
(Proof)	G.	934	59.1			G. 29	8.6	3.0			0.35	
11	B.	924	54.8	1: 5.1	− 0.08 ± 0.02	B. 12	8.3	3.1	60	0.18 ± 0.02	0.37	0.88
(Proof)	G.	921	59.8			G. 5	7.3	3.5			0.48	
12	B.	827	66.1	1: 6,540	+ 0.17 ± 0.02	B. 73	11.0	3.0	56	0.15 ± 0.02	0.28	0.72
(Proof)	G.	1021	55.8			G. 64	10.2	4.0			0.39	
13	B.	1063	15.2	1: 163	− 0.13 ± 0.02	B. 15	4.3	4.2	59	0.32 ± 0.04	0.38	1.09
(Proof)	G.	1485	20.8			G. 2	2.6	2.4			0.90	